MW00886604

Two Into The Blue

Two Into The Blue

The Story of The Gemini Launch Program Told
By a Man who Participated in Every Launch

Robert L. Adcock

Copyright © 2009 by Robert L. Adcock.

Library of Congress Control Number: 2008905265
ISBN: Hardcover   978-1-4363-5021-1
                   Softcover   978-1-4363-5020-4

All rights reserved. No part of this book may be reproduced or transmitted in
any form or by any means, electronic or mechanical, including photocopying,
recording, or by any information storage and retrieval system, without permission
in writing from the copyright owner.

This book was printed in the United States of America.

To order additional copies of this book, contact:
Xlibris Corporation
1-888-795-4274
www.Xlibris.com
Orders@Xlibris.com
50032

CONTENTS

Dedication

This book is dedicated in memory of all the Martin Company employees, for their tireless efforts in bringing about this successful program. Those in the field, at the launch site, who are too many to name, will forever occupy a favorite part of my memory. We had a common objective that required unselfish dedication on the part of every person to get the job done.

I want to thank Justin Bryant for his aid in review and grammatical correction and Jim Kelley for his review and technical editing. Thanks to other contributors: Wally Feagan, Ed McMechen, and Chuck Cicchetti for their program remembrances.

Preface

G emini was the precursor to the Apollo lunar program. The technology of rendezvous and docking, so vital to Apollo, was perfected. Gemini encompassed ten manned missions. Along with solving the problems encountered with extravehicular activities (EVA), the program accomplished a host of secondary objectives and kept the astronaut corps flying, ad interim, while Apollo was still under development.

This story is about the National Aeronautics and Space Administration Gemini program written from the personal point of view of someone who worked directly helping launch the astronauts from the Cape Canaveral Air Force Station when space travel was in its infancy. Other books documenting the program have provided a wider, more programmatic view, rightly focusing on the astronauts and their accomplishments. Here I have tried to focus on the prelaunch activities of the launch vehicle and the spacecraft because I was present for, and had a significant job with, each of the twelve launches. The Gemini program was named after the Gemini stellar constellation containing two stars, Castor and Pollux, symbolic for the two astronauts who rode each flight.

The United States was in a space race with the Union of Soviet Socialist Republics (USSR), and until the time of this program, the USSR was more technically advanced. With Ed White's EVA on Gemini IV, the United States began to narrow the technical margins with Russia in space accomplishments.

The launch team was a select group of experienced engineers and technicians from the Titan II program, and they demonstrated it in the success the Gemini enjoyed throughout the program. We were young (probably all less than forty) and motivated by the desire to do our jobs. After all, no one in this country had ever launched *two* astronauts into orbit at the same time, especially on a Titan II intercontinental ballistic missile (ICBM). The first concern of the program was for pilot safety, and this ranged in all aspects from design and manufacturing of the flight hardware through all its prelaunch activities at Cape Canaveral. Redundancy was designed into the space vehicle systems, manufacturing was accomplished under rigid-control conditions, and prelaunch processing was governed by a set of rigorous procedures designed to ferret out any flaws and preclude failure in any hardware or software.

Many other contractors[1] and government agencies were involved in the program. Those who worked with the Gemini hardware and were responsible for launching it benefited by previous-launch-experience technology developed on predecessor programs.

---

[1] Other major contractors providing flight hardware other than mentioned herein were Advanced Technology Laboratories, spacecraft horizon sensors; Bell Aerosystems Co., Agena propulsion; General Dynamics for the Atlas; Lockheed Missiles and Space Company for the Agena target vehicle; General Electric, guidance and tracking; Honeywell Inc., guidance components and systems; IBM, flight computers; Motorola Inc., spacecraft command systems; North American Aviation Inc., OAMS and RCS; and Northrup, parachutes.

Robert L. Adcock

Chapter 1

View from the Top

I arrived, along with the other launch team members, at launch complex 19 at 3:45 a.m. on June 3, 1965. Complex 19 was the Cape Canaveral Station launch area assigned to NASA's two-man Gemini Space Program. That day we were going to launch the Gemini IV space vehicle, manned by astronauts Jim McDivitt and Ed White. This was to be the second manned launch of the program, and the astronauts were to try two new things: rendezvous with the spent launch vehicle second stage in orbit right after it separated from the spacecraft and an EVA (extravehicular activity—a space walk outside the spacecraft).

Gemini came after Mercury, America's first manned space program. Gemini's principal objectives were developing rendezvous and docking techniques, training a cadre of astronauts for Apollo and developing EVA procedures. At this time, there had been two unmanned Gemini launches. Only six weeks earlier, the first manned launch of this program, with astronauts Gus Grissom and John Young aboard, had flown from this complex. It was now up to McDivitt and White to make major progress on Gemini IV toward some of the program objectives.

The air of expectation at the launchpad was electric that morning as launch personnel began arriving at work. The fresh guys just coming in were relieving their counterparts, who had been there all the previous night. Busy Range Operations people setting up film cameras, searchlight operators checking out their lights, and a host of others talking on radios with speakers blaring—all contributed to the cacophony in the otherwise quiet, warm Florida night. Strange electric lights indicating status of unidentified equipment sparkled from the edges of the palmetto bushes outside the enclosed launch complex. The equipment, in most cases, turned out to be portable remote film camera equipment that had been brought out earlier. This gear allowed special movie cameras to start—some at a preset time in the count and others at an event such as vehicle liftoff. A whole army of people came out of the woodwork on launch day and provided a host of services, from guard services to in-flight tracking of the vehicle. Most of these services were provided by the Range Operations contractors.

The security guard contingent that night at the launch complex access gate had been doubled to deal with checking out the extra people going into and out of the blockhouse area. We needed a special badge to get into the blockhouse and launchpad area. To get inside the gate, you had to trade your Cape Canaveral access badge for the one that indicated you had been through the background check. The guard kept these badges on pegs at the gate badge board. At that time of night, at shift change, portable floodlights had been set up over the badge board so the guards could match the workers' faces to the badge pictures. The diesel generators that powered these lights clacked away, contributing to the overall din. Some of the guards' supervisors were standing by, and the squawking security guard car radios sounded ominously throughout the night and early morning. All these temporary preparations and activities gave a surrealistic quality to an otherwise familiar workplace scene.

On the Cape proper, outside the launch complex, the broadcast media hype had been homing in on the launch all week. America had not yet launched too many men into space, and public interest was running high for this flight. The news media sent their reporters down to the Cape days in advance of the launch. They wandered around, preparing their press releases, hungry for a story; and they talked to whomever they could about the upcoming launch. Occasionally, the US Air Force let them talk to some of the contractor workers. Jules Bergman of ABC TV became a regular visitor to the pad. Walter Cronkite was somewhere in the area. *Gannett News* interviewed me for an article that ran in the local paper because I was "somebody important" on that launch day.

The main Martin Company manager at Gemini Launch Complex 19 was Frank Carey. Frank was called chief test conductor and supervisor of Martin Launch Operations. Every hands-on (those who touched or operated the hardware) Martin employee assigned at the launch complex worked for him. Launch operations was the hands and arms of Martin's Gemini effort in Florida. They were the only ones who operated the Titan hardware and launched it.

Test conductors led and coordinated the activities of each shift in preparing and launching the vehicle. When a new launch vehicle arrived at the launch complex, a test conductor and his crew would be assigned to follow that vehicle through its processing, count it down, and launch. Most launch cycles, from receipt to launch, took about two months to accomplish. I was the test conductor assigned to Gemini-Titan IV. My crew and I had been working on Gemini-Titan since it arrived from Baltimore—the Martin factory—throughout the checkout. It was now my responsibility to be in charge of the launch vehicle through its launch-countdown operation on the day that McDivitt and White flew.

Jim Kelley, the night-shift test conductor, met me in the blockhouse control room that morning and gave me a "good news" greeting. He and his team had loaded the propellants the night before. The most difficult aspect of the previous night's work schedule was the launch vehicle propellant loading. That was always an *iffy* four- to six-hour operation. The propellants are highly toxic to humans and, for that reason, cumbersome to handle. The amounts that were loaded into each tank had to be within very strict tolerances. Beginning with the first Gemini launch vehicle preparations, propellant loading always occupied a significant part of the prelaunch preparations. What started out as a simple tanking test, because of its notoriety and frequency, later became known as the WETS (the Wednesday Evening Tanking Society). Later when combined with spacecraft to launch vehicle verification, it became part of the test, Wet Mock Simulated Launch (WMSL). Even later we realized that a more realistic simulation would be to load the propellants for the WMSL the night before, so that the WMSL became a two-day test, and propellants were loaded not under the stress of a countdown clock. Of all the tests, propellant loading or tanking was always performed at least once for each launch vehicle.

Jim's was good news going into the test conductor's meeting. I got a cup of coffee and we reviewed the work status from the evening before and our readiness to start the launch countdown. Our customer was the US Air Force. NASA obtained the Titan from the US Air Force, who contracted with Martin to build it for use in lifting the Gemini spacecraft into low earth orbit. The Titan was designed as an Air Force intercontinental ballistic missile weapon. Therefore, the Air Force was involved heavily in this peaceful use of the Titan.

I chaired the test conductor's meeting that was held in the ready room conference room. Frank Carey, the Lead engineers, and the Quality Assurance supervision, all from Martin, attended as well as McDonnell, Range, and other parties. The meeting objective was to review the readiness for flight. The Air Force contingent on site managing the launch operations was the 6555th Test Group of the Systems Command. Both NASA and the 6555th had their own employees at this meeting, and each had their own versions of how well things had fared the previous evening. There were a lot of other interested parties in attendance. McDonnell spacecraft personnel also attended. The room got pretty crowded even at this early hour. First, we reviewed with the customer any open items that had been written during the evening or that had been left over from earlier. An item was a notation that was recorded if a piece of hardware or a procedure failed to meet specification, and it was open if that failure still needed resolution. The Quality Assurance people recorded the open-items list, and they were the only ones who could close them. They would not close an item on their own initiative. There had to be a corrective action specified, and successfully carried out, before an item could be closed. I expected, as did everybody else, that by this time in the processing cycle, all the open items would have been closed; and they usually were. At this meeting all GT-4 open items were cleared, and we got a "GO" from NASA and the 6555th to begin the countdown.

We got out of the meeting, proceeded to the blockhouse control room, and started the countdown on time. The countdown proceeded normally to T-100 minutes when McDivitt and White entered the spacecraft. All the work on the spacecraft, which included assisting the astronauts getting onto their flight couches (seats), was carried out by McDonnell Aircraft people. They too had to load propellants, install ordnance, and do a host of electronic checkouts—although more intricate, but usually in smaller quantities than on Titan. A NASA test conductor(TC) managed all these spacecraft-related activities and the NASA/Martin interface testing between the Titan and the spacecraft. The NASA TC sat in the blockhouse control room adjacent to the Titan TC, and that enhanced communication between them.

Events in the countdown continued on schedule for the rest of the count until T-35 minutes when it was time to lower the erector. The erector was a movable tower that got its name from the Titan program. On Gemini, the erector was used in the same way as for Titan; and when it stood vertical, it allowed people access to the launch vehicle and the spacecraft, totally enclosing the stack where the complete vehicle stood on the launch pad mount. When, during the countdown, all work was done on both the Titan and the Gemini spacecraft, and access to the flight hardware was no longer needed, the erector was lowered to lie flat on the ground, out of the way of the blast of the rocket engines at liftoff. T-35 minutes was the erector-lowering time in the countdown; and when it was down, that was the last event on the pad before launch, whereupon everybody cleared back to the blockhouse. From that point on until launch, the astronauts laid in the spacecraft and were the only ones on the pad. The only way for them to get out of their spacecraft at this time was to raise the erector back to vertical and get them out normally as they had entered or, in an emergency such as a fire, to eject using their rocket-propelled seats.

On this day, the erector wouldn't go lower than an angle of about forty-five degrees above the horizontal. Jim Houghton, the Martin mechanical lead engineer, was sitting at the erector-control console in the blockhouse; and he called to let me know they were having a problem. They fiddled with the problem for a few minutes—Houghton working with his guys who were still located at the lowering machinery on the pad. I called a *hold* at T-35 minutes until we could get this problem fixed. Houghton soon became frustrated with the progress of the problem resolution, and he announced he was going out to the pad. Frank jumped up and went with him.

Meanwhile, somebody outside the immediate launch area suggested the mechanical people at the pad might need some more expertise. Frank, my boss, knew that an engineer named Wally Feagan probably could be a lot of assistance in troubleshooting the problem. He had worked the previous shift and was probably at home in Mims, ten miles north of Titusville. It would be very difficult to drive a car from Mims down US Highway 1 to the Cape because the crowds for launch viewing had traffic choked for miles (there was no Interstate 95 back then). Mr. Feagan got a call from Frank who told him he was needed and to come on in to work. By the time he got dressed and got down as far as Titusville, he heard on his car radio that we had worked around the issue and resumed the countdown. He then turned the car around and went home.

With Frank and my lead engineer, Houghton, still at the pad and both too engrossed with the problem solving to talk, I was left in the dark about the details of what was happening on the erector. Other things needed attention on the Titan, and I was mildly busy with them during this hold. Naturally, the customer and NASA, along with the astronauts, wanted to know when it would be fixed and when we could get on with the launch. Christopher Kraft (the flight director) called me from Mission Control in Houston, asking about the status of the troubleshooting. He wanted to know what was going on and when we could get going again. I amazed myself with the answer I gave. I summarized the work effort, saying we had the best, most knowledgeable people at the pad working on it; and as for the time when we could get going again, well, the standard answer when you didn't know for sure was to estimate ten more minutes. He let me off easy because he surely surmised that I didn't know much more about the problem than he did (he was surrounded by experts at Houston who were advising him).

Finally, after about an hour, I got a report that the problem had been overcome, and the erector started down to its fully lowered

Robert L. Adcock

position. How the issue was resolved was that at the pad, some techs sawed the handles off two brooms and used the handles to hold in the electrical power contactor that caused the lowering motors to run! Frank came back in, and we started the count again after a seventy-six-minute-hold. I didn't have time for many questions, and both he and I were busy getting the Titan systems going again to pick up the launch countdown.

From about T-120 seconds and down, things got pretty busy on the launch vehicle. We had to do things like arm the destruct system ordnance, start the launch vehicle hydraulic pumps used to guide the rocket in flight, open valves to let the propellants flow down to the engines, and transfer the launch vehicle to internal battery power. There we were; cameras were running, the public was watching from the viewing areas and on TV, and I was sitting on the edge of my chair, waiting for the ground launch sequencer to click off the seconds and hoping nobody would have a reason to scream "hold" in my ear! The two astronauts in the spacecraft atop this potentially deadly mixture of aluminum, steel, propellants, and ordnance—waiting for the ride of their lives—were more or less blasé about the goings-on in the blockhouse. They were models of patience. Some astronauts, it was rumored, had even dozed off in the final minutes of the count, waiting for the last of the preparations to be completed!

All eyes were fixed on the time posted by the countdown readout units installed inside the blockhouse. At T-10 seconds, our announcer, Jerry Walden, his voice ringing out in its nasal tones, started counting backward over the public address system and the voice network: ten . . . nine . . . eight . . . seven . . . six . . . five . . . four . . . three . . . two . . . one. When the count reached zero, two propellant valves on the Titan opened; and the two individual liquid propellants, driven by a turbopump, gushed together in the first stage engine thrust chambers, igniting on contact (these are hypergolic fuels, meaning they ignite on contact with each other). Burning furiously, the propellants started producing the engine thrust buildup until it reached full value, four hundred and thirty thousand pounds for almost two seconds. Having sustained full thrust for that time, the ground launch sequencer exploded four bolts that held the space vehicle stack to the pad, and the flight began. The Titan's engines ran at full thrust until the first-stage propellants were burned. The engines had no throttle as the shuttle has today. It was analogous to starting your automobile engine, laying a concrete block on the accelerator, pulling the transmission into drive, and letting go of the brake for a two-and-a-half minute run at full blast down the road. It was flat out for better than two minutes.

If you could have been at the launchpad during this launch sequence, it would have seemed like a monstrous cyclone of fire, metal, vapors, and steam swirling and exploding and sounding like a screaming banshee. The tiny gyros and electronics high up in the front end of the Titan sensed the vehicle's every movement and made order out of this mayhem. When the bolts were exploded and set the vehicle free, any abnormal movement such as the vehicle starting to tip over was immediately corrected. You can get some idea of the control problem by trying to balance a broomstick in the air with the tip of your finger: it's unstable, and you continually have to correct it to keep it from falling. It rose slowly, gaining speed as the engines gulped propellants, and continued pushing on an ever-lighter load.

When the sequencer reached zero, I watched the shutdown light to make sure it didn't light; then the roar of the engines through the blockhouse's twelve-foot-thick walls confirmed they were running, and about that time, the green liftoff light came on. A quick glance at our only closed-circuit TV screen in the blockhouse confirmed at a glimpse that the rocket flew past the umbilical tower!

We launched at 10:16 AM. I remarked to Frank after liftoff that the dwell time of the vehicle in flight before it cleared the pad umbilical tower seemed unusually long. He remarked that they all seemed that way to him! He was responsible to the astronauts for the safety of their flight until after they had cleared the launchpad umbilical tower! He announced over the net, "Tower clear."

I was elated! I had never sat in the test conductor's seat before at the start of a manned expedition. In a way, the feeling was like when I first soloed an airplane during flight lessons: I knew I could do it, but things happened so fast it was over before it even got underway, or so it seemed. I had joined a very exclusive club. Only a handful of men had done this job before—launching two men atop a dangerous machine and having them survive! I felt emotionally very close to each one of the launch team; I wanted to be with them, to celebrate, to have a beer, and to share our experiences. I needed to talk over the events, to relive the critical moments of the countdown with those who had just lived through it! This was heady stuff for a Tennessee country boy, and I was enjoying it.

After the launch, we all left work, and some of us stopped off at Bernard's Surf Restaurant in Cocoa Beach. Since the early-launch days, the Surf was where a lot of different personnel from various companies hanged out and traded stories from work. Bernard Fisher, the owner, was enthusiastic about the space effort. His employees could anticipate when a crowd was coming in from the Cape, and they poured good drinks! When we first got there, Frank got a phone call from his boss, Joe Verlander, who told him he had to go back to work! Apparently, Mr. Tibbs, the Martin Company Canaveral division vice president, disliked the outcome of the erector problem; and he was livid. He called an emergency conference at his office.

One of the directors wanted to fire Wally Feagan, the guy Frank called in on the erector problem because there was some question about his maintenance procedures on the machinery. Mr. Tibbs wanted to do something positive to preclude this debacle from happening again, so at the suggestion of another director, he gave the responsibility for future erector operations to Jack Hagan, the Martin director of quality assurance. The operating responsibility was taken away from Martin Launch Operations (our group). This was unheard in operations circles: nobody gave quality assurance any operations job! Mr. Hagan, at that point, requested an engineer who had expertise to help him with this new assignment.

The director of engineering suggested Wally Feagan as the most knowledgeable engineer about this equipment in Martin. The irony of the story is that at one minute, Wally was the goat, and the next he was a hero who practically had his job guaranteed for the life of the program! Henceforth when we needed to operate the erector, we called the erector directors, Hagan and Feagan.

Meanwhile, the Gemini IV powered flight lasted a little longer than six minutes, as planned. Its flight plans included the task of station keeping with the second stage of the launch vehicle just after it had been jettisoned. Strobe lights had been installed in the second stage so the astronauts could find this target. Station keeping proved to be tricky, and after they used about half of the fuel in the Orbit Attitude Maneuver System (OAMS), Houston called the experiment off. They needed to preserve the OAMS fuel for use during reentry. The crew estimated the closest they got to the orbiting second stage was two miles (about 3.2 kilometers), but in so doing, the astronauts demonstrated that they could make the significant orbital plane change maneuvers needed for rendezvous and docking.

Ed White performed the first American EVA on Gemini IV. Aleksey A. Leonov, a Russian cosmonaut, had done an EVA in March 1965. The Russians' recent EVA gave an incentive to Ed to do his EVA earlier in the program than planned. Imagine the thrill of Ed stepping out into space, some 103 miles (166 kilometers) above earth! On the third orbit pass, White moved outside the spacecraft. He had a handheld maneuvering gun he used to position himself around the spacecraft. The propellant in this gun was soon exhausted, and White then maneuvered via his tether around the spacecraft, positioning himself to shoot photographs of the spacecraft and earth. Originally, the EVA was scheduled to last twelve minutes. He stayed outside longer than planned due partly to his euphoria derived from the view and due also to the time it took to get all his gear back in and stowed for splashdown. When he was finally talked back in, he exclaimed, "It's the saddest moment of my life." The EVA showed that man could perform useful work outside a spaceship—a capability most useful in the latter parts of the Apollo program.

McDivitt and White had been assigned as the prime crew for Gemini IV in July 1964, and the backup crew were Frank Borman and James Lovell Jr. They were allowed to observe the manufacturing and individual tailoring of their spacecraft. For example, the flight seats were individually made for each astronaut, as were their protective suits. Thus, when they flew, they already had almost a year of experience with their hardware and with the Houston Flight Operations people who would support their flight.[1]

---

[1] Hacker, Barton C. And Grimwood, James M., *On the Shoulders of Titans*: National Aeronautics and Space Administration, 1977, page 240.

Prior to GT-4, NASA Flight Control was located at Cape Canaveral; but by the time of Gemini IV, they moved the Mission Control Center to Houston quarters. The move was completed in time to support Gemini IV. Also, for the first time, the planned duration of Gemini IV was long enough that Mission Control team members had to be on duty around the clock to provide coverage for that flight operation.[2]

The initial plan had been for Gemini IV to be a seven-day flight if electric power was available. That plan included a practice rendezvous using the rendezvous evaluation pod installed on the spacecraft since there was no rendezvous vehicle. For the first time, electric power was to have been generated by fuel cells. Later, during the spacecraft assembly and development, when fuel cell delivery lagged, planning reduced the mission to four days so that it could be accomplished using only batteries for electrical power. The EVA had been planned for a later flight, but equipment and training became available early, so it was moved up and incorporated into this flight plan.[3]

Extending the in-orbit duration from the three revolutions experienced on Gemini III to four days planned for Gemini IV concerned the medical community. They argued that four days of weightlessness followed by the sudden experience of being back to ground level in a full gravity environment could cause the astronauts cardiac problems. The astronauts were in superb physical shape, and they would be exercising while in orbit by using a bungee cord device rigged for the exercise regime. Despite these concerns, NASA bit the bullet and went ahead with the flight. It was a good bet because subsequent postflight examinations of the crew revealed that although there had been some weight loss, some loss of bone mass, and some loss of blood volume, they were in good physical shape.[4]

Of the other major experiments carried on by Gemini IV, photography was one of the outstanding ones. In synoptic terrain photography, the crew took photos of well-known earth's terrain

---

[2]   *Titans*, Pages 239,253.

[3]   *Ibid.*

[4]   *Ibid.* page 253.

features. These photographs would later serve as standards with which to evaluate lesser-known earth areas.[5]

At 12:44 PM (EDT), cruising above Hawaii after sixty-two orbital passes, Gemini IV started reentry. About thirty minutes later the spacecraft hit the Atlantic Ocean, forty-eight miles (seventy-seven kilometers) away from the *Wasp* aircraft carrier there to pick them up. The ship sent a helicopter to the point of spacecraft impact. The helicopter raised the astronauts up, then winched the spacecraft out of the water. With the astronauts safely aboard the *Wasp* and the spacecraft sitting on the carrier deck, the flight of Gemini IV ended.[6]

[5] *Ibid.*

[6] *Ibid.*

Robert L. Adcock

Chapter 2

What's a Gemini?

Six years after starting with Martin, I was working at Vandenberg Air Force Base, California, on Titan II, an intercontinental ballistic missile (ICBM) weapon system the Air Force was developing. Our family was growing, and we decided to buy a house because with two children, we just weren't making it in the apartment we had lived in the last two years. We had just come back from a temporary three-month assignment in Denver, where I had been sent to learn about the Titan II ground-support equipment, and we needed the new house right away. We convinced the seller that he should let us move in immediately even though we hadn't closed on the house. Meanwhile, I had written my former boss and friend in Baltimore, Bob Schlechter, who previously was the Vanguard project director at Canaveral. I asked him about the possibility of coming back to Baltimore to work for him. I also knew that would be an almost sure way to make it back to Cape Canaveral; that is, if I started work on a Baltimore project, I might then be transferred with it to the Cape.

I liked working in Florida. The technical atmosphere in Florida seemed more like a laboratory environment where people were more *gung ho* about getting their jobs done. That feeling was in my blood, and I wanted to return to it. I had almost forgotten about my letter to Bob in Baltimore, but no sooner than we had moved into the newly bought house, his answer came back. He said he would assign me to the US Air Force Dyna-Soar program as soon as I could get there. That was good enough for me, and I began to see how we could get out of the house deal. I found the owner would gladly keep my earnest money deposit and let us break the contract. Even though we loved that house, the prospect of getting back to Baltimore and a new adventure in our lives thrilled us, so we transferred in November 1961.

I enjoyed working with the people on the Dyna-Soar program. I had started my tenure with Martin in Baltimore, so this was my second time living there. I recognized many familiar faces from the first time I had been there. I had a good feel for the new project, and things started going well at work. The community we settled in, Lutherville, right outside Towson, was a good bedroom community. Bob Schlechter, my boss, was one of the finest and smartest gentlemen I have ever met.

The Dyna-Soar was a US Air Force Air Research and Development Command project. It was a manned program that was to boost astronauts into orbit. They would orbit for a preplanned time and glide back to earth under pilot control, much like the shuttle *dead-stick* reentry today. As the program developed, the weight and mission demands of Dyna-Soar had driven the launch-lifting requirements to a new Titan III booster, which was essentially a Titan II with solid rocket motor strap-ons, thereby enormously increasing the throw-weight capability. Martin's role in all this was to design and build the Titan III and be the overall program integrator. Although the Titan III was being engineered in the Denver division, Martin in Baltimore had the program-integration role, which was the part I was working on.

Martin bid on NASA's Apollo Command/Service Module and on the Lunar Excursion Module. Shortly before Christmas 1961, Martin announced both projects had been won by the competition. A number of jobs were lost at Martin when neither of these projects materialized.

Shortly after the Christmas break, a Dyna-Soar engineering administrator came by my desk and told my colleagues and me, "Put all our working papers in his cardboard box so we can turn it in to the government." We even turned in yellow pads with no more than doodles and phone numbers on them. We asked if the documents were to be identified and put into some sort of classified system so that whoever got them could make heads or tails of them. His answer was no; we were to turn them in all in a box, and that was the end of Dyna-Soar as far as we in Baltimore were concerned.

Of course, everyone went home on time that night, and I must have told my wife about the sudden end to the project, but I don't remember us worrying much about my job. The next day, my little test group was sitting around when the same administrator came in with some time cards and said, "Charge your time to this, *Gemini*."

"What's a *Gemini*?" we all asked in unison. It wasn't long before we had it figured out. My boss had a meeting and gave all his guys the general program information. We were told that the Martin Company would produce the launch vehicle, basically a Titan II; and McDonnell would build the spacecraft, a Mercury spacecraft that was modified to accommodate two astronauts. Martin Denver would fabricate the propellant tanks because they were building the Titan II weapon system, and these tanks would be carbon copies of the weapon-system tanks. We, in Baltimore, would bolt the four tanks together and add the electronics, hydraulics, engines, and interstage structure—all designed and built in Baltimore (except for the engines and radio guidance systems that were built by others, in factories elsewhere and shipped to the Martin assembly facility).

The Gemini-Titan II was quite a diversion from the weapon-system version because of man-rating the launch vehicle. *Man-rating* meant beefing up or adding redundancy to systems where a flight hardware failure would cause mission failure and jeopardize the lives of the astronauts. Thus, redundancy was built into the guidance and flight control systems, and a new system for detection of hardware malfunctions was added to warn the astronauts of an in-flight failure.

The Titan II was a two-stage booster. One engine with two thrust chambers on the first stage generated 430,000 pounds of thrust. Aerojet General Inc. built the engines at Sacramento, California. During flight, the burnout of the first stage came about two and one-half minutes after liftoff after all the propellants were burned; and when that happened, the second-stage engine ignited. By separating and dropping the first stage and getting rid of its dead weight, the second stage could proceed on in flight with its reduced thrust of about one hundred thousand pounds. The second stage and the bolted-on spacecraft separated from the first stage and flew on alone until both the stage 2 and the spacecraft reached orbital velocity. Both went into orbit. At that time, the spacecraft separated from the second stage, and using its own thrusters adjusted its velocity to that which was required to achieve their initial orbit. The second stage of the launch vehicle was not controlled after burnout and therefore tumbled about in space. Increased atmospheric drag caused by its tumbling made it reenter the atmosphere and burn up.

As mentioned, man-rated had resulted in redundant flight control and guidance systems. This amounted to independent guidance and control system strings. At the head of each string was a guidance system whose steering commands were issued to the autopilots that translated these commands into corresponding engine movements powered by hydraulic actuators both on the first- and second-stage engines. The primary string was a radio guidance system produced by General Electric, and the secondary was an inertial guidance system in the spacecraft built by Honeywell Inc. Two separate electronic autopilots and attendant systems completed two redundant, independent flight-control strings. Either system could be used to get to orbit, but a flight rule required that both systems were working and in the primary mode (the radio guidance string in that case) before liftoff. Additionally, a malfunction detection system was added to detect flight malfunctions (if any) early on, before they became catastrophic, and provide warning to the astronauts in time for them to escape.

I was assigned to the Martin Baltimore Test Group that defined the total test program for the Gemini launch vehicle, especially everything that was to be done before it was shipped to Florida. The company wanted to sell (that is, get a government form DD250 that signified ownership of the vehicle by the government) the launch vehicle to the US Air Force to show that it was working OK and ready to start launch preps when it left the factory in Baltimore bound for Florida. This precedent was established in Denver on the Titan II, and because the Air Force would transport the vehicle to the field by aircraft, it was a good time for the government to own it.

Dyna-Soar had a similar sell-off planned at the Baltimore plant. During that program, a test cell in Martin's C building was under construction. There, the vehicle would have stood erect and the launchpad, insofar as the vehicle electronics and other systems were concerned, would have been simulated. We would have done subsystem checkouts and integrated tests and sold off the vehicle there. We just couldn't have loaded propellants or ordnance or launched. The test cell was called the Vertical Test Fixture (VTF). It was a fixture because it was provided to the project as a test tool. The Air Force owned it. A similar postassembly block of tests and sell-off was fashioned for the Gemini launch vehicle.

The largest part of my assignment was to accomplish the launch vehicle vertical test and sell-off in the VTF. My job was to get the people onboard who would actually run the VTF and get the fixture activated. The construction work on the fixture was already started, the steel erected, and some of the work platforms in place when the Gemini program began. We had to get the Aerospace Ground Equipment (AGE) installed, tested, and ready to receive the launch vehicle when manufacturing was complete in D building. It wasn't long before problems began to come up at the program manager's meetings that affected the VTF; and Bob Schlechter told Mr. Bastian C. Hello, the Gemini program Martin director, that I could handle the problems for him. Suddenly, I was in Mr. Hello's office often to brief him on VTF status because vertical testing was on the critical path schedule before shipping the vehicle to Florida. One thing about reporting directly to the program manager was that things got done when someone else on the project held up the action! I must have bothered him a lot in the early days. One day I came into his office with a new problem, and he exclaimed, "Adcock, every time I turn over a rock, I find you under there!"

One Saturday morning before we were almost done activating the VTF, Mr. Jerry LaFrance, the Martin Gemini chief design engineer, came down to the VTF and wanted me to show him just what we were going to do with the launch vehicle when we got it in the VTF. I gave him the tour and the story about it, and when we returned to my cubicle in the control center, I showed him a schedule-type chart I had made of our planned testing activities along with constraints and estimated span times for each activity. I explained that it had been my experience in this sort of thing that much of the time we had allocated to the test cell would be spent waiting for someone, somewhere, to get us something we needed to complete a job. In our daily planning, and using my chart, we planned to look ahead to all the events; and if everything that was needed for that test was on hand, we would stick a green pin in that bubble. If, however, something was missing—e.g., major hardware—and it had a promised delivery date that would not support the test schedule, then we stuck a red pin in this planned activity. Tests that were completed were left with green pins inserted.

Some of the program managers who worked for Mr. Hello were responsible for getting us the things to the VTF to support the test schedule. It might be, I explained, Design Engineering that owed us some drawing that would clear up a constraint, or it might be Logistics who owed us a part or maybe Manufacturing who owed us a subassembly. So we took the picture of the manager whose organization was delinquent for that activity, and we displayed that picture on our display board as "man of the week" for everyone to see. Mr. La France was quite impressed, and he apparently told Mr. Hello, who implemented a carbon copy of our chart in his office. Our planning people undertook to update the chart in his office so it would reflect what was in mine. I was told that the discipline managers worked hard to ensure that their pictures were not on the "manager of the week" board in Mr. Hello's office! When Mr. Hello had a weekly staff meeting, they reviewed who the "man of the week" was and why.

The men I had already working for me doing planning were top-notch guys; but we still had to acquire a few good men for the dedicated test team who would reside in the VTF. I thought of Will Thackston in Vandenberg. He was just finishing up the Titan project there. I had known Will from the Vanguard program at the Cape a few years before. I wanted to give the VTF a launch-site flavor rather than a design-development test flavor. We hired Harry McCaw as our electrical lead engineer, also out of Vandenberg, and for propulsion we selected Ray Navin. Al Ryan headed up the guidance and control group, and Tony Fragomeni led the instrumentation group. Ed Chapman was lead for MDS and Range Safety systems. Other outstanding guys were Al Schuchardt of Range Safety and Jim Moore of guidance and control. I wanted the VTF people to be independent from the design engineers so they could view the product from the same perspective as the Martin people at the Canaveral launch site. Bob Schlechter agreed with this approach. I explained this to Will, and he did a great job in setting this up and getting the people trained and organized.

I had an agreement with Bob Schlechter that once the first vehicle was sold off, I could accompany it to Florida. I still hadn't shaken the sand out of my shoes. I wanted to get back to the Cape. My wife was reared on Merritt Island, and her folks were still there. I thought also that if there was anything I was good at, it was probably launching rockets. I worked a deal with Lenny Arnowitz, my former boss on the Vanguard program, to move me to Florida in the Titan III Engineering organization that he was building The plan was for Will Thackston to take over my job at the VTF when I left and continue to run it for the remainder of the program.

In those days, we were working to the program schedule that had the first Gemini launch in December 1963. That meant that we had to be done with VTF testing and acceptance sometime around early August 1963 so we could ship and do all the testing at the Cape in time to support this date. When the GLV-1 arrived in June at the VTF from the manufacturing line in the factory, just about every system was missing at least one component. This was probably not too unusual for the first article of its kind of any assembly line in almost any industry, whether it be automobiles or tractors or whatever. Authorized modifications were yet to be worked in some instances. Nevertheless, we applied electrical power to the vehicle on schedule and started subsystem checkout. For a first-time vehicle checkout, things went very well for the first of many subsystem tests until hardware shortages and wrong configuration hardware started to have an impact on the schedule. It began to look bad for making the schedule to support a first launch date in December.

Eventually, all the missing hardware and paper came together, allowing the VTF to finish the launch vehicle checkout. Suddenly, the focus was on the VTF people to get the vehicle tested and shipped to the Cape. One thing that helped was that several of the Martin launch team members from Florida visited the VTF and helped out in areas where we were shorthanded. They were training themselves and helping us at the same time.

We were so behind schedule that we worked on the Fourth of July that year. Yet we didn't have much design engineering support that day. The program manager, Mr. Hello, helped us out down to the VTF. He parked himself at Forest Caldwell's desk and held an ongoing operations review throughout the day. We would bring up one problem after another. He would ask us what and who we needed, then ask loudly, "Where is that guy?" Most often they had taken off for the Fourth, and he would get the guy on the phone and scream for him to come in and support us. As he reared back in the desk chair talking on the phone with one foot on the desk, he was kicking the top of the desk as if to punctuate what he was saying—*thump . . . thump . . . thump*! He helped put the importance of the VTF in perspective for a lot of people.

On July 31, 1963, just twenty-seven days later, we started Combined System Testing (CSAT). The first, an electromagnetic interference test, showed several areas of interference between subsystems. Modifications to the ground and airborne systems were effected to reduce these problems so that by September 5, all interference problems were solved. Following that, the CSAT was completed, and the data were presented to the Air Force and the Aerospace Inc. acceptance team on September 11. After about a week of looking at the data and inspecting manufacturing records, a physical inspection of the vehicle revealed hairline cracks in the inserts of some of the airborne electrical connectors. This started a physical inspection with the inspectors using flashlights with ten-power magnifying glasses on all the connectors. We found several bad connector inserts. A broken idler gear in one of the Aerojet engine turbo pumps (shipped to Baltimore that way), plus the fact that forty-two airborne components were not yet certified for flight (that is, they hadn't completed development-qualification testing) caused the acceptance team to reject the vehicle. Rework, removal, and replacement of the damaged connectors and associated wiring took until October 2. The formal second run of the second CSAT was completed on October 4. The vehicle acceptance team reconvened and accepted GLV-1 two days later and gave permission to ship it to the Cape on October 26, 1963.

Robert L. Adcock

After completing the first CSAT, I felt it would be OK to go ahead with my move to Florida. The CSAT had gone well despite the problems that held up the acceptance. I had set a date with the mover to pick up my household goods. Then at work we found out about the connector problem, but the boss felt it safe for me to go ahead with my move before full acceptance of the first launch vehicle. I left town as the VTF gang was working on connector replacement. I was happy to be going to Florida but uneasy about leaving Baltimore before the work on GLV-1 was finished.

Chapter 3

Back to Florida

My family and I took some vacation time when driving from Baltimore to the Cocoa Beach area, and we stopped along the way to see my relatives in Alabama. Mom had not seen the grandkids in some time—in fact, she had never seen Carol, the youngest (she would have been about six months old at the time). We stayed there two days. When we left, my wife drove the larger car with Carol, the new baby, and Robert, the middle child. I was driving the Volkswagen with Rebecca, the eldest child. We had been on the road for close to ten days. The cars were loaded with pillows, quilts, and blankets and all kinds of children's toys. It must have been some sight because my mother remarked to my sister as we were leaving, going out their drive, "Looks like the 'hind wheels' of bad luck going there!"

When we got to Florida, we visited with Barbara's dad a few days in Titusville while looking for a place to live. We had lived in this area before, but now everything was bustling, caused primarily by the activity at the Cape and at Kennedy Space Center. When we left in 1959, there was no Kennedy Space Center; and now shortly after the president's announcement about going to the moon, North Merritt Island was a flurry of activity: men and machines carving up the natural lay of the land to make way for the launching and return of astronauts from the moon. After the president's death in 1963, it was dubbed Kennedy Space Center; but before that, I believe it was called the NASA Launch Operations Center.

Since we now had three children, we were looking for a nice place that could accommodate the five of us. We found a furnished place at the Saturn Apartments in Cocoa Beach. It was pretty nice having a place to hang our hats after having been on the road for a long time. I was slated to report to the Titan III program at the Cape. Frank Carey wanted to borrow me, so I wound up going over to the Gemini Launch Complex 19 on a temporary basis to wait for the GLV-1 arrival from Baltimore.

I reported to Frank at Launch Complex 19, and he gave me a desk in the test conductor's office in the ready room building, just outside the main gate. Frank's plan was to have two regular work shifts of engineers and technicians, both launch capable and supervised by a test conductor. When I arrived, there were in place two test conductors, Charles "Chuck" Cicchetti and James A. Kelley. Markus Goodkind was an assistant test conductor. I soon learned the problems they faced in getting the launch complex ready to launch the first Gemini mission. The launch vehicle had not yet arrived, and the Martin people were struggling to finish the activation testing and complete writing all the test procedures that would be used in checking out the launch vehicle.

The USAF provided the launchpad for Gemini. It was a Titan I pad, Complex 19, and already existed because it was built to launch the Titan I weapon system. Since Titan I was already deployed, Launch Complex 19 was declared surplus and made available to support the Gemini program. The 6555th Test Group, under the direction of Colonel John J. Albert and responsible to the Space System Division of the USAF Systems Command in Los Angeles assumed the role for modifying and activating the launch complex and supervising the launch vehicle contractor in day-to-day operations. Martin was their prime contractor in getting the modification work done through the Martin activations group, under the direction of Peter Abate. The Gemini activation work was essentially complete when I first arrived at the Cape in early October 1963. At this time the launch operations people had assumed operation responsibility from the activations group and were moving into the launchpad facilities.

I had met Frank Carey on the Titan I program in Florida when I came down to visit from Vandenberg the year before I was assigned to Gemini. As I recall, he was a flight control engineer at that time. During my days in Baltimore while checking out GLV-1, I was vaguely aware of the activities in Florida concerning the Martin problems, getting the ground hardware to work and the launchpad modifications and activation done. Of course, we had our own set of problems in Baltimore dealing with the launch vehicle. A Florida launch team was being assembled that would conduct the launchpad activation testing, and then go on to become the vehicle launch team. I knew Frank had been selected as chief test conductor early on, and one of his assignments was to select the test engineers and technicians who would do the work. Most of these people I did not know. Frank welcomed me with open arms when I arrived because I had intimate knowledge of the vehicle after my experiences in Baltimore for the past one hundred days or so.

Most of the team had served with Frank on the Titan I. They came with good reputations based on their long run of Titan launch successes. On Gemini, however, a policy was implemented to train and certify people both individually and as a team no matter what their previous experience was. This was new, and to each individual, it meant going to class to learn the systems, passing exams on the material, and then being "stand-boarded." *Stand-boarding* meant the individual had to appear before a panel of experts on the system and answer questions about how the system operated. Some saw it as an insult to their intelligence or a black mark against their professionalism. Every person who would touch the flight hardware in any way, including the quality inspectors, had to go through this process. We are talking roughly two hundred people, and in the aggregate, a lot of manpower was expended in training and certification. It was done because Martin was going to fly men for the first time, and management wanted nothing left to chance that could cause a catastrophe and potential injury to the astronauts.

When operations began on the flight vehicle and launch teams for each work shift were designated, the business of stand-boarding the teams began. On the launch vehicle, we had enough men so that individual operations could be carried on each shift, and either shift team could support the launch. We had, in effect, two individual launch teams. The approach was to take a team whose members had already been individually certified, and in a simulated operation, technical problems would be introduced in the system; and the team as a whole was graded on its ability to take action to save the vehicle and isolate the problem. On Gemini, we used the actual launch vehicle connected to its ground equipment to simulate the malfunctions (today, simulators made for this purpose are a matter of course; but in 1963, we had no such machines). We would get the team members in countdown posture on station, with the launch vehicle systems operating, just as if we were in a launch countdown. Unbeknownst to the operating people, a problem would be introduced into the ground equipment (the testers could do this simply by lifting a wire or flipping a switch outside the view of the operators), and to the operators in the blockhouse, it looked like something had really happened to the launch vehicle or the spacecraft. It was realistic enough to make you sweat! The Air Force personnel and their Aerospace Corporation counterparts were usually looking on during these simulations, and although they said very little, we all thought they were probably making judgments about the individuals involved in these exercises. They were nervous about us using the hardware for these simulations, but with no other way to simulate the launch, they reluctantly accepted these exercises. They were afraid of damage we might do that would go undetected. It was a learning experience.

In my own case, my certification was not yet complete, although I had extensive training and experience on the launch vehicle back in Baltimore. At the launch site, several other major interfaces existed that required training. A test conductor needed to understand these interfaces fully, and one such interface was with the Range. The Range's role during launch was so large that it took two major contractors to perform the function for the USAF. They were Pan American World Airways (PAA) and Radio Corporation of America (RCA). PAA handled the building's maintenance, roads, services such as air-conditioning, food service, and guard services, while RCA took care of photography, in-flight tracking, range safety, telephones, and ground electronics. What this meant is that Pan Am and RCA had people scattered all over the Cape and at the island tracking stations. These were huge contracts involving thousands of people. Supporting their people and getting them coordinated to support the launches was a logistical nightmare! Somehow these people had to communicate with one another and their coordinator/supervisors during their launch countdown support. This involved all kinds of radio and telephone communications, even ship-to-shore, for the tracking ships. Satellite communications were in their infancy and not yet used in those days. I think they did a superb job. The launch vehicle interface with the Range during a launch countdown was handled through the Range-supplied superintendent of Range Operations (SRO) who coordinated all Range activity in support of the launch. More SRO stories will be told later!

One of the major concerns for any launch operation was public safety. PAA and RCA provided the data necessary for the Range safety officer (a USAF officer) to decide whether to let the vehicle continue to fly if it deviated from its preplanned course. The Range safety officer always had the option of terminating the flight, which meant he had radio cutoff commands to the flight engines. Once they were shut down, he could then proceed to blow up the vehicle if he thought people on the ground were endangered! Such an action would have been tough on astronauts flying on the vehicle; so on Gemini, there was about a three-second time delay between engine shutdown and destruct, allowing the astronauts to abort the mission. There was a perpetual disagreement between the flight system developers and the Range Safety people about the hazards of carrying explosives to destroy the vehicles that had astronauts aboard and whether the astronauts could have override power for the explosives or have them only in the early parts of the flight; disable them later or not to have them at all. In the end, Range Safety always prevailed and had this destruction capability which was never used on any of the Gemini flights. Data upon which this destruct decision depended involved all kinds of radar and visual tracking during the early powered flight and also vehicle performance data throughout the flight. On the islands downrange there were tracking stations and at least six stations were located at Cape Canaveral and the islands of Grand Bahama, Eleuthera, San Salvador, Grand Turk, and Antigua. A tracking ship was also added to the network for gathering in-flight data, called the *Range Tracker*. This constituted the Range-tracking network used primarily for Range Safety, as opposed to the NASA flight-tracking network that supported the spacecraft flight in real time.[1] The people who worked with these two Range contractors were doing all kinds of work in support of any launch, mainly gathering in-flight data and retransmitting it back to Cape Canaveral Central Control.

The superintendent of Range operations (SRO) sat in the Central Control building and communicated with all Range contractor personnel during a countdown. On one of the Titan II launch attempts, I got to go over there and sit through a countdown with the SRO. Martin's people on another launch site were trying to launch their weapon system missile that day, and I got to sit through this launch attempt.

I met George Small, the SRO assigned to the Titan II launch that day. He had four other guys all in a room without windows, sitting at desks behind banks of phones, wearing headsets and talking furiously to different people. I didn't know what they were doing at first, but I began to figure it out the longer I sat there. From the time I got there until it was over, a stream of information about things going wrong (or in some cases, going right) was pouring into these men, and they were reporting all to George. He was so cool! He was on the headset to the test conductor in the blockhouse at launchpad 16 where they were counting down the missile, and to that test conductor, everything on the Range was peachy keen. But to the Range people there seemed to be nothing but problems. One of the major problems in that operation was a tracking radar mounted on a ship out in the Atlantic that had been inundated by rain the night before and was out of order. The ship's crew was trying to get it dried out and working before it was needed in the flight. They had an hour or two until they checked it out with

---

[1]   For a complete description of both the flight tracking stations see Grimwood, James M., Hacker, Barton C. and Vorzimmer, Peter J. *Project Gemini, A Chronology:* NASA Report SP-4002, appendix 4, page 282.

the missile during the count. Regular reports were coming in on their progress or lack of it. Finally, when the count "started to get serious" as we said in the business, meaning when the countdown got down to just minutes away from launch, the SRO needed an answer from the ship. George got on the phone with the ship's captain and talked to him like a Dutch uncle until they both decided the Range could not support this launch with that radar out. As I recall, the Range reported a "NO-GO" when the Complex 16 test conductor polled everyone at about T-5 minutes, and then they scrubbed the launch.

But that wasn't all that happened that day. While the SRO was wrestling with the ship's radar problem, all kinds of other odd things were happening. I remember one with clarity. There was a "destruct command" transmitter (a large device that resembled a radar) at the south end of the Cape that could have been used to destroy the vehicle in flight. Naturally, the transmitter was checked in the early part of the countdown, together with the missile's response, to ensure perfect working order. That day the people running the transmitter blew an electrical fuse, and they used up their spares. The time of day when they called the SRO for help was in the rush hour when southbound traffic was its heaviest. Knowing that no truck could possibly go from the warehouse to the transmitter in time to support the launch, they suggested a novel approach to the SRO. Get a spare fuse and give it to the PAA security police who, with sirens and lights, could clear the way to the transmitter site in time for the scheduled checkout. I came away from this experience with total respect for Range Operations. Whenever I could, I always requested George Small on their end during our countdowns.

One of my early assignments was working on the procedure for the Combined System Test (CST). The CST was an electrical simulation of the launch vehicle's launch and flight. No propellants were loaded, and no ordnance was onboard during this test. The CST was very much like the CSAT in Baltimore's VTF, and I copied as much as I could from that document when writing this procedure. From the CST procedure, it was a short hop to write the launch countdown procedure. What a great training aid this assignment became! Not only did one need to know the launch vehicle systems, but we also needed to understand these unique ground systems interfaces that were unavailable to us in Baltimore. Some were unique to the launchpad, and I had to get involved with the spacecraft people for the first time to get their spacecraft part of this document.

One of these interfaces for the CST in Florida, at the pad—that was different from Baltimore—was the airborne radio guidance system would be commanded by the General Electric ground station on the Cape called MOD III. GE had their idea on how it was going to command the airborne guidance (ours) from the MOD III ground (theirs) during CST, and it was contrary to what we at Martin wanted. It boiled up to an issue, and some of our design folks from Baltimore came down to help settle it. It escalated up the chain to the 6555th and NASA for resolution, and that was when I first met NASA's Walt Kapyran. In a meeting with NASA, Air Force, Aerospace, GE, and Martin and after a few questions by Kapyran, he ruled in favor of GE and we went on from there to do it the GE way. We were afraid the ground station could do damage to the launch vehicle displacement gyros if GE did what they said, but there was never any damage.

The spacecraft to launch vehicle electrical interface was one thoroughly tested because of its complexity. Even before the mechanical portion was completely mated, we did an Electrical Interface Integrated Validation and Joint Guidance and Control Test. This tested all the control systems that made up the redundant guidance-flight control systems, i.e., the primary radio guidance flight control string and the secondary inertial guidance string. The inertial guidance package was in the spacecraft but drove the secondary autopilot and remainder of that string located in the launch vehicle.

The final electrical test of this interface came just before launch preparations began. It was called the Joint Combined System Test in which both the spacecraft and the launch vehicle simulated a countdown and abbreviated flight.

The activation of Complex 19 was essentially complete, but a few things still needed cleanup. One of them was the propellant loading system. Valves in the system were a problem (faulty manufacturing of valve seats, which meant they couldn't be exposed to the toxic propellants without being damaged); and at some point early on, all the valves had to be removed out of the system and returned to the vendor for new seats installation. The ground system test had to be completed and then a demonstration that the system could load the propellants onto the launch vehicle. We were in the throes of testing this system when someone came in the blockhouse and said he had heard on the radio that President John F. Kennedy had been shot to death in Dallas. Later, we took off for the weekend, and my family and I headed to Orlando to spend Thanksgiving with Dave Mackey and his family. Dave and I had been friends since the Vanguard program, and we hadn't see each other since I came back to Florida. Dave worked for Martin, Orlando. We watched all the memorial services and news that weekend with Dave and his family.

The launch vehicle had successfully passed many tests in Baltimore. Aside from the propellant problems mentioned above, all the other major interfaces with the spacecraft, the Range, and other ground systems at the launchpad still had to be demonstrated. We did a one-time sequence compatibility firing to show that the airborne systems would cycle through their sequences and fire both the first- and second-stage engines while the launch vehicle was being held down on the launchpad. We also did an electromagnetic compatibility test to ascertain that the unique ground equipment in Florida did not interfere with the functioning of the airborne equipment. These tests, because they involved all the launch vehicle systems, took up a lot of schedule time. It became obvious that we could not launch in 1963 because too many things were on the launch team's plate.

One of these things was the modification for the launch vehicle pogo problem. Pogo was a phenomenon that caused the launch vehicle to oscillate along its longitudinal axis during powered flight. It stemmed from the reaction of the accelerating vehicle when it reached the area of maximum dynamic air pressure encountered at about eighty seconds after liftoff. The system, so configured, became unstable in this part of the flight regime and produced these unwanted oscillations. If left unfixed during a Gemini flight, it would shake the astronauts to the point where they would have a hard time functioning. The vehicle acted like a kid on a pogo stick that bounced up and down on the ground when the stick is jumped on. Although this phenomenon was OK for the weapon system, it would give the astronauts a rough, shaky ride. NASA specified just how much shaking the astronauts could stand without causing them a problem. The launch vehicle change had to be made in Florida by the launch team before the first flight to prove its effectiveness even though it was an unmanned flight.

Before the launch team could start to make a hardware change, the Martin Company project people and engineers got together with the customer and got their authorization for the change, designed the change, and manufactured the hardware that was to be installed on the launch vehicle. This entire package—authorization, engineering drawings, and parts—showed up as a kit at the launchpad for the Martin Launch Operations people to install. They got to work and continued on into Saturday night when a slight mechanical interference occurred between the new hardware "steerhorns" and the stage 1 engine turbo pumps, i.e., their outsides rubbed together. The *steerhorns* were dynamic fluid dampers installed in the first-stage propellant feed lines between the tanks and the engine. The modification engineering authorized the filing of a small part of the engine turbo pump fin so that it wouldn't interfere any longer, if required. This didn't meet with the local Air Force approval, however, and some strong words passed between the Martin test conductor in charge on that shift and the local Air Force guy at the pad. Martin's position was that the mod was authorized, the engineering was available, and the people knew what they were doing; so they were proceeding. The Air Force opinion was that they were not buying a fix that required filing off parts of the engine. Inasmuch as it was Saturday night, higher authority was hard to locate at home to secure resolution to this impasse. The Air Force representative at the launchpad decided they were in charge and ordered the Pan American guards to evacuate the launchpad and lock the gate until Monday morning! This may have been another *first* for the program. There had not been another case that anyone could remember where the contractor was locked out from the launchpad.

The first spacecraft, which was just a "boilerplate" model with weights inside to simulate the real spacecraft, was brought to the launchpad in February 1964 to mate with the launch vehicle. The first flight was just to show that the Titan could put a spacecraft of that weight in the right orbit and that all the Titan modifications worked. Several tests were run to show compatibility of these two major parts before launch.

The launch countdown was conducted on April 8, 1964. In that count, I was assigned as launchpad control reporting directly to Chuck Cicchetti, the launch vehicle test conductor in the blockhouse. I was his eyes and ears out at the pad. I reported on the status of all work going on at the pad area and on the vehicle. The spacecraft people were out of sight in their white room that surrounded the spacecraft, at the top of the erector, but I could see them coming and going because they had to board the elevator at the base of the vehicle where I stood. There were two elevators—one went up to the individual levels at the launch vehicle access positions, and the other went directly to the spacecraft white room and was used by the astronauts and other spacecraft workers for direct access to their hardware.

The count started some six to seven hours before flight; and the things we had to do were the final checkouts of electronic systems, final checks with the Range for range safety systems, final connection of all ordnance, folding and stowing all erector access platforms, and securing the lanyard cables that disconnect the umbilical lines from the vehicle at liftoff. All these had to done before we could lower the erector to the launch position. The erector folded like a jackknife and laid down flat on the launchpad surface during the launch. We were scheduled to lower it at T-35 minutes in the count. Early on, getting close to the T-35-minute mark, I noticed that a lot of work remained to be done: folding work platforms, and getting everything ready to lower it. All the mechanics who were supposed to be doing these things were busy folding the spacecraft white room up like a pasteboard box so it would clear the vehicle when the erector was lowered. We needed more help at that time! I was about ready to tell the test conductor we couldn't get ready for erector lowering on time and we would have to hold the count when, Jim Kelley—another one of the test conductors—walked up, and I explained the problem to him. He immediately went down to where the Propulsion people were standing by, and suddenly I looked and there were about four or five additional people starting to fold access platforms and secure lines helping get ready for erector lowering. They were all over that erector at every level!

Robert L. Adcock

Jim later reported, "I went to the propulsion room, the electrical room, and the instrumentation room and got the service of five technicians with tethered hammers and crescent wrenches, and we all climbed up the ladder to the sixth level and started working. I motivated the techs by telling them that I would authorize two hours of overtime which would be added to each one's time card." At T-38 minutes I called the test conductor and told him the pad was ready for step 1 at T-35 minutes (erector lowering), and the TC responded, "Go, mother." This had been a critical time, and Jim had saved the day for us!

As the erector was being lowered, one of the electrical umbilical cable lanyards got hung up on a platform folded segment when it was about seventy-five degrees from the horizontal and threatened to damage the umbilical if we continued the lowering. I thought we would have to return the erector back to the vertical to fix this problem when Wally Feagan, the shift mechanical lead engineer (the same guy in chapter 1 who became one of the erector directors) at the pad, climbed the erector, like a monkey scaling a tree, and dislodged the lanyard. He didn't even need an elevator! I am sure he violated all kinds of safety policies by doing that, but he sure saved the day.

That was the last thing we had to do at the pad before retiring to the blockhouse for the launch. I later asked Jim Kelley how he knew that those Propulsion technicians could get the erector ready for lowering, and he told me that those people were once mechanical techs on a different pad and on a different missile program. He knew them and knew they were well versed in how the erector worked.

The launch went off well. The Titan carried the spacecraft into a two-hundred-mile orbit. The spacecraft did not separate from the second stage—it wasn't supposed to—and the combination stayed in that orbit until April 12 (a duration of four days) when it reentered and burned up. This was as expected. It was the project's first launch, and it proved so many things, mainly the Titan's capability and sound design.

After the launch, the test conductor and I felt energized, and we went to Cocoa Beach in the motel area looking for a party. We went first to the Colony Seven Motel that was named by the original seven Mercury astronauts. We thought this surely would be the place where the party action was. We saw one room where a few people whom we knew were coming and going, and we also started in. The guy at the door took one look at us and recognized we were Martin employees and said, "Get out of here, you bums, this is for the VIPs."

We retreated from this room, and while we were outside recovering from this rebuff, we looked across the street to the Holiday Inn and spotted a large gathering of people there. We went over and joined them. We didn't know who the crowd was, but we knew it was a party! A line formed leading into one of the rooms, so we got in-line. Soon we began to recognize some of the McDonnell spacecraft people (from pad 19 blockhouse), and we stood in-line talking with them. As the line progressed and I came into the room, suddenly I was personally greeted by the owner and president of McDonnell Aircraft himself, James S. McDonnell! I was a little chagrinned to admit I worked for Martin (not McDonnell) to which he replied that he once had worked for Glenn L. Martin before organizing his own company. We chatted for a couple more minutes, and he said for us to have a drink on him, and he appreciated our efforts of that day.

The launch had gone extremely well. It was a shot in the arm for the program and everybody on it! Next day we knew we had to get busy because despite the euphoria from this launch, the program was nevertheless behind the original schedule by about seven months. The people at the launch site would be expected to make up some of the lost time and get all the launches done on schedule.

Chapter 4

Letting Go

Management sometimes wondered why the vehicle had to stay on the pad so long before it flew. Why did we need to put so much manpower into the prelaunch preparations? The answer was among other things that all the obvious installations had to be accomplished, such as installing the spacecraft, the ordnance, modifications and the propellants. There was also a minimum set of tests to be performed to prove the systems were working correctly and that they would probably continue to work throughout the upcoming flight. Once it left the ground, there was no ability to adjust the system should it develop a quirk in flight. It had to leave the ground with the best probability of working throughout the flight by giving it all the love and care you could on the ground. With the added responsibility of two astronauts on top of the rocket in the equation, the ground team had quite a chore to decide if it was OK to launch, meaning let it go! We accomplished this by emphasis on the people we chose to work there, by their training, certification and motivation; by paying attention to the details of the hardware performance during checkout; and by imbuing those closest to the systems with accountability and authority for flight safety.

The launch vehicle work—i.e., checking out, servicing, and launching—was done by the launch operations group of Martin. The work was divided up according to the individual launch vehicle systems. On Gemini, those were Electrical, Mechanical, Propulsion,

Instrumentation, and Radio Frequency systems. A lead engineer was assigned to each system, and he in turn had one or more system test engineers and technicians who were his hands-on people working with the hardware. The Electrical people under the direction of Tom Wirth took care of the rocket electrical power systems, the in-flight sequencing systems, the flight control, and hydraulic systems.

Other systems were equally complex. For example, Propulsion managed by Ken Shipe (later by Vern Derby) had the stage 1 and 2 engines, all the plumbing and valves, pressurization system, and controls. Given the complexity of the airborne systems and the myriad of ground equipment needed to checkout, operate, and service the systems, the lead engineer's responsibility was enormous.

Don Striby (later Lou Favata) managed the Instrumentation system that contained all the airborne telemetry measurements as well as the ground system instrumentation including blockhouse readouts and recorders.

Jay Sain was responsible for the Radio Frequency systems that included the Malfunction Detection, the Radio Guidance, and the Command Control and Destruct.

Jim Houghton had the Mechanical systems including the erector, the airborne vehicle structure, and air-conditioning. These lead engineers were carefully chosen for their jobs because of their technical maturity and administrative abilities.

The lead engineers took ownership of their assigned systems. They didn't want anyone but their people "messing around with it." They controlled the configuration of it, and their people ran all the tests and operated the systems. The lead had more knowledge about the health of his system than anybody. The tests were designed to show the system performed to its specifications, and unless it met specs, it wouldn't fly. Just about all systems had to meet military specifications that were numbered thusly: MIL-E-XXXX, where *MIL-E* stood for military electrical specification with serial number XXXX, for example. Then there was a whole hierarchy of contractor specifications, a subset of the military specifications, and drawings that spelled out the specific system requirements and configurations.

There once was a joke about field guys modifying and repairing the flight hardware. The joke went, all work done in the field should meet a unique military spec called MIL-F-41CE, which meant,"Make it like the frigging engineering for once!"The program folks in Baltimore considered the Florida people "the field." The best in the field you could do was to make the hardware's form, fit, or function like it was designed by the design organization. Sometimes we uncovered designers' mistakes, but rarely.

The Martin lead operations engineers lived with the launch vehicle hardware and its operation. Because of that, they checked its health daily and could have noticed something in passing that might have escaped the untrained eye. The system lead could veto the decision to fly based on his gut feeling. Few did, but nobody would usurp this authority from him. Without their say-so, the launch operations people would not have felt free to *let go* and launch the mission.

The launch operations lead engineers were responsible to a test conductor (TC) who had been assigned for that flight. Test conductors were responsible for a flight's technical excellence. The TC had as much information about the systems as his lead engineers, and they spent time daily exchanging information on each system checkout status and health. The test conductor was responsible for conducting all multisystem tests, such as the launch countdown. He was in charge of all launchpad activities in that countdown on that day and managed all integrated tests involving the launch vehicle, spacecraft, and range operations. The test conductor was responsible to the Air Force and NASA customers for recommending a launch and had to prove his recommendation through reviews of paperwork and open items. The customer, of course, would have been following the checkout activities daily; and when they came to the system status review meetings, they were themselves already prepared for what they were about to hear.

The test conductor had a venerable reputation in the aerospace industry. I remember early on in the Vanguard days, Bob Schlechter would go around, randomly asking the launch team employees during the countdown on launch day who they worked for. According to Bob, the right answer was,"We all work for the test conductor on this day, including myself, because getting the launch off on time and safely is our biggest job today."

With that kind of reverence and awe associated with the job, I knew I wanted to be a test conductor someday. On Gemini, I was to be one of the few who had manned launch responsibility; and in retrospect, I did get to be launch conductor on vehicles 4, 8, and 11.

The launch countdown operation was the culmination of all the prelaunch work. So many tasks had to happen at the last minute before the launch it was natural that they would be organized on a time-based sequence so that the test results were unambiguous. This also ensured that there would not be any interference between tasks.

Most of the Gemini launch vehicles differed only slightly from one another only because of unique hardware needed to accomplish a particular assigned mission.

On our team, it was kind of an unwritten rule that the test conductor, who was planning to conduct that countdown and launch, would write that countdown procedure and develop that vehicle's test and checkout schedule. It was up to him to know everything in the procedure document and why it was in there. That meant he had to know the planned activities not only of the launch vehicle but also of the spacecraft and the Range. The countdown procedure was rewritten for each new mission launch because flight objectives changed. When the document was rewritten, it was reapproved by both NASA and the Air Force. Our countdown document was the one official document that governed all launchpad activities on launch day, no matter who was doing them whether it was the launch vehicle people, spacecraft, or Range. There were subsidiary documents these organizational entities used to perform their work, but they were integrated with the "Master Launch Vehicle Countdown" document. Everyone agreed on that document because it was their organization's commitment to do the things called out for them and when and where they were to do them. Once approved, it was distributed to the world. The main players were NASA and their contractors, the Range and its contractors, and our customer, the Air Force and Martin.

The major jobs the launch vehicle people had to do on the launch day were to turn on and check out all the electronic systems, check out the Range and NASA interfaces with the airborne electronic systems, check the electronic interfaces with the spacecraft, connect ordnance, pressurize the propellant tanks to flight pressures, and lower the erector. The toxic rocket propellants would have already been loaded into the tanks the evening before, and the space between the liquids and the top of the tanks, called the ullage, was filled with low-pressure nitrogen gas. Flight pressurization meant increasing the ullage pressures to twenty to thirty pounds approximately per square inch, depending on which tank was being pressurized. The spacecraft people would have had to do the same kind of work, including loading the astronaut crew into the spacecraft. Their tasks had to be coordinated with the launch vehicle so as not to interfere with one another's work. The countdown document coordinated all these tasks.

The interface with the Range was enormous. Electronically, the Range received launch vehicle telemetry and recorded it. They also tracked the vehicle in flight to make sure it didn't deviate from its planned flight path. Range Flight Safety had the ability to terminate the flight during the vehicle's powered flight. The destruct checks with the Range's destruct transmitter that were made during the countdown had to be conducted before the live destruct initiators could be electrically connected.

The Range also provided tracking of the launch vehicle flight using an airborne beacon. Certain interlocks on the automatic launch sequencer (Range provided) also had to be checked so that Range Safety could stop the countdown and block the launch if their requirements were not met—for example, if the Range was not clear of boats and people when we were ready for launch. All of these interfaces would have already been checked out before launch day, but the launch-day checks were to ensure they were still working before starting the final phases of the count and flight.

One of those tests designed to test the flight hardware in its flight configuration with the Range and with NASA Mission Control interfaces was called the Flight Readiness Test (FRT). The FRT was very close in hardware configuration to that on the launch day. This test was done a few days before the launch countdown was initiated. The configuration of the hardware during the FRT was, the spacecraft was installed on the launch vehicle; the astronauts were in the spacecraft in a pure oxygen environment, simulating the conditions at liftoff; the erector was lowered; and no propellants or ordnance was aboard. The Range and NASA tracking groups could interrogate the spacecraft and launch vehicle radio frequency systems open loop (through open air) without checkout antennas. The FRT was built upon an abbreviated countdown that ran from about T-30 minutes to the point in plus time where the time of spacecraft separation from the launch vehicle in orbit would have occurred. This was the only time the Range and Mission ops personnel could look at the flight vehicle open loop (standing alone), in flight configuration, with electronics free to radiate, before they had to do it late in the countdown on launch day. So it was a good and necessary test.

In the launch countdown, certain tasks, like ordnance connections, required radio frequency silence. For safety reasons, all the workers, except a bare minimum, remained inside the blockhouse. It soon becomes apparent how incompatible, from a safety standpoint, some of the jobs were with one another. We had it arranged so that all the hazardous work—the arming of the destruct system, the connection of all other ordnance, and the pressurization of propellant tanks to flight pressures—was done before the astronauts were allowed to get in the spacecraft. In that way, we assured that their exposure to hazardous operations was minimized.

The early part of the Gemini-launch countdown had to proceed to a preselected point, then *hold* until the desired and actual liftoff times optimized the conditions aloft for effective rendezvous. The built-in hold of nominally about five minutes' duration (meaning the hold point was preplanned and written into the countdown document) length was adjustable, and the sooner the rendezvous was to take place in orbit after the launch, the shorter the hold time. For example, this length of time was less than a minute on mission 11 in order for them to rendezvous with the target on its first orbital pass. So the coordination of everyone had to be superb in order to meet these launch constraints.

On the manned Gemini program, the approach to making the launch vehicle launch decision was incumbent primarily on the

Robert L. Adcock

field launch operations people. Martin's system engineers and quality-control personnel (looking over the shoulders of their launch operations people) assured the vehicle really did have the best chance to fly correctly. So on any one operation or test, there were launch operations people performing the test, quality control people witnessing the test and buying off the procedural steps as they were performed, system design engineers looking at data from the test, and sometimes government quality control people too. In a large test such as propellant loading, people were at their assigned stations both in the blockhouse and at the propellant farms. Everyone communicated with one another over voice headsets. As test conductor for this operation, we ensured we had a Range Safety "GO" to start propellants flowing to the vehicle and that all needed support was on hand. We then turned the actual flow operation over to the Propulsion engineer who directed the techs on station at the farm. It was always comforting to me when we asked a tech to read a meter and report, he would say, for example, "Two thousand gallons," immediately followed by a second voice saying, "Two verify," meaning the quality guy verified the tech's reading! That was a source of comfort, knowing that another set of eyes had looked and saw the same thing! But successful test completion depended not only on the ground test hardware giving an OK and the test observers signifying their approval, but also on a formal after-the-fact meeting in which these same personnel reviewed the completed test procedures and the test data.

A lot of effort went into ensuring that the hardware was ready to fly. The final "GO" for launch did not occur until a few days before launch in a meeting called the Flight Readiness Review (FRR). In that meeting, higher management reviewed all the completed procedures. They reviewed all the problem reports and the corrective actions and, in all cases of airborne components failure, the formal analyses of these failed components. The FRR had all the contractors in attendance together with the Range, the USAF, NASA, and the astronauts. Sometimes the FRR would last for more than one full day. From that day forward to launch day, we reviewed with the customer and one another all open items that had been written since the FRR, and we had to explain how these would be closed before launch.

Hardware was also designed with redundancy to increase its system reliability. The first-stage hydraulic system, the flight control autopilots, gyros, and guidance systems were redundant. Critical circuits in the launch vehicle were also redundant. Forty-six major components in the launch vehicle were classified as flight *critical*. A component was judged to be flight critical if its failure in flight could abort the mission or cause injury to the astronauts. Critical components required special handling, special storage, and a pedigreed history of their life span. Failure of a critical component during checkout required a formal failure analysis and report, explaining how and why it failed. These failures were reviewed time and again during the posttest critiques and during the FRR.

One unique approach that may have, by now, become an industry standard was the approach of having a "missile mother"—a person who was assigned to the hardware when the manufacturing first began. This person followed the hardware everywhere it went, from station to station, location to location, until it left the ground. This meant that for the launch vehicle, the first place a missile mother needed to be was in Denver, where they welded the vehicle tanks. His travels then took him to Baltimore when the tanks were shipped for final launch vehicle assembly. Ultimately, he wound up at the Cape, where it was launched. All aerospace hardware had a traveling log that listed everything done to the hardware, but the missile mother's job was to remember everything that ever happened to the hardware, from construction through launch. They became walking encyclopedias concerning the vehicle, relying on notes and reports. Their instant recall of things that happened in the past to that hardware was invaluable. There were four of these individuals, and they followed all the launch vehicles through their launches.

George Taylor managed the Quality Assurance group at the launchpad, and his supervisors were Paul Glynn and Ed McMechen. They were charged with keeping records of things that had happened on the vehicle during its life here on earth, and they presented these data to the customer before every flight. Their organization was the sole official source of these data presented in the Flight Readiness Review

Robert L. Adcock

before each flight. Getting past a Flight Readiness Review was a major hurdle in the preparations for the launch countdown.

It was because of these built-in safeguards on the Gemini that we didn't have too much of a problem letting go when the time came. We had made it as close to MIL-F-41CE as possible!

Chapter 5

Acts of God

Gemini launch vehicle 2 made it through the factory testing and sell-off much faster than GLV-1. Since GLV-1 was still on the launchpad, GLV-2 was kept in Baltimore to incorporate modifications before sending it to the Cape. It arrived and was installed at Complex 19 by July 11, 1964. Subsystem testing began with the expectation that it would be complete and the launch vehicle would be ready to mate to spacecraft in mid-August. That was in plenty of time, but the spacecraft was still in St. Louis due to parts shortage and waiting for completion of testing.

A thunderstorm passed the Launch Complex 19 area on Monday evening, August 17. Although there was no direct lightning hit on the launch vehicle, later testing began to indicate failures in several of the airborne flight systems. With so many failures showing up in different systems at the same time, we wondered if they had been caused by an electromagnetic phenomenon triggered by the thunderstorm. When we understood the full extent of the damage to the failed Titan subsystems, we decided, just to be on the safe side, to replace all launch vehicle electronics packages. Solid-state circuitry was fairly new in those days, and nobody wanted to take a chance that some circuit had been damaged and would fail in flight. This meant we had to start all subsystem testing over again because we replaced everything! Apparently the lightning strike was close enough to the launch complex to cause an electromagnetic disturbance of sufficient magnitude to overstress some of the solid-state components in all the launch vehicle airborne packages. It happened at night when almost nobody was at work. Normally we kept one man there in off-shift hours, and the man with the duty that night had not observed anything out of the ordinary but the storm.

Before we could get over the lightning retesting, Hurricane Cleo threatened a direct hit on Cape Canaveral. Management decided to remove the launch vehicle from the pad and put it in storage out of the weather. Frank Carey called me at home on the morning as the storm was blowing in. I was outside at my home in Cocoa Beach, trying to put up some plywood over my sliding-glass doors, and already the wind was howling. Frank asked me to get together a Mechanical Team to take the launch vehicle down and store it in a hangar out of the storm. The spacecraft had not yet been installed and was out of harm's way. I was able to get about five other men, and we met at Complex 19. All we had to do was reverse the erection procedure by taking the second stage down first followed by the first stage. The erector played an important part in this procedure. We placed handling rings around the second stage. These were tied off to the erector. When the bolts holding stage 2 to stage 1 were removed, the second stage was floating inside the erector, suspended by cables. Then we lowered the erector with stage 2 still suspended by cables inside it. When it was horizontal, we drove the second-stage transporter, a wheeled, rubber-tired vehicle, under the stage. We loaded and secured the stage on the transporter with both still inside the erector and then backed the transporter out with stage 2 still on it and towed it to the storage hangar.

By the time we got everything together and got started, the wind had to be gusting to fifty miles per hour, but the gear used to secure the second stage to the erector was strong, and we didn't have any problem with it. When we got ready to start to take down stage 1, the storm had grown in intensity to where it was not safe to be handling it. I didn't quite know what to do—I must have called Frank back, but I also talked to Joe Verlander, Martin's Gemini project director. The decision was made to let the first stage ride out the storm standing vertically on the pad with the erector lowered, exposed to the elements! It was feared that if the erector was left up, relative movement between the erector and the vehicle, caused by high wind, might damage the vehicle. The Titan lived up to its name—it was built strong like a *brick guardhouse*! So we sat out the rest of the hurricane at our homes and left the first stage standing on the launchpad without the erector around it while the second stage was securely in a hangar back in the industrial area.

When testing had barely resumed following this hurricane, on September 8, a new hurricane, Dora, appeared to be heading directly for the Cape; and again, Martin Company launch team personnel took down the rocket and stored it away. Dora passed by, missing the Cape, but by that time, of all things, Hurricane Ethel was threatening, and reinstallation of the rocket back on the pad was delayed until September 14. The spacecraft arrived in Florida on September 21 and was brought to the launchpad on October 18. The effect of all these unplanned events put us further behind in the schedule. It squashed any NASA plans that the first manned flight would be in 1964. We would first have to launch vehicle GT-2 and then go through vehicle GT-3 processing, and there just wasn't time before the end of the year.

Gemini 2's launch countdown started on December 8. Jim Kelley was the Martin test conductor, and I was assigned the pad control. The pad control's job was to watch and direct the activities at the launchpad until everyone returned to the blockhouse for launch. You would think his job would be done when everybody was off the pad and the blockhouse door shut for launch. Not on Gemini. The pad control came back to the blockhouse and stood by in the firing room in his SCAPE underwear, looking out at the pad through a periscope like those used in submarines. The term SCAPE stood for Self-Contained Atmospheric Protective Ensemble. The purpose of being in the underwear, rather than regular street clothes, was that if we had to get to the pad in a hurry in SCAPE, then we were already half-dressed. It was humiliating to be there in the control room in front of all those other men dressed only in long johns! The long johns-type underwear was to wick up perspiration when we had donned the SCAPE suit. Due to the toxicity of the Titan propellants, all personnel had to be dressed in SCAPE if we had to return to the launchpad this late in the count. In the unlikely event of a propellant leak, we were totally protected by the SCAPE for thirty minutes—that's how long the air supply lasted in the suit.

This was our first countdown in which the spacecraft was active although unmanned. It had flight systems installed. One of these was the reentry control system (RCS), used to position the spacecraft in the correct orbital attitude for atmospheric reentry when starting to return to earth. I guess NASA spacecraft upper management was nervous about whether or not the system was working, so they decided to fire the RCS thrusters late in the countdown to give them assurance. Jim Kelley, the launch vehicle TC said, "George Page, the NASA spacecraft test conductor, was directed to repeat the RCS firing at T-20 minutes. I (the launch vehicle TC) had witnessed an RCS firing before, and it took about twelve minutes after the firing to reconfigure the system for launch. I told the launch sequencer technician (who was in charge of running the countdown sequencer) that at T-10 minutes, 'I would be busy trying to get a "GO" from the spacecraft TC, Page(SIC-Page).' I did not say anything to him, I stated to him to hold the launch count at T-5:30 minutes and recycle to T-20." Jim was sitting next to George Page in the blockhouse, and he continued, "At T-9 minutes, I asked Page on and off the network for a spacecraft "GO" for launch, and he stated off net that his RCS system was too busy preparing for launch. At T-5:30 minutes, the (countdown) clock stopped, and I called (all launch) personnel to recycle to T-20 minutes and standby until I received a "GO" from George."[1]

I'm standing there in my underwear, and as I recall, everyone seemed surprised we held, but the cause of the hold was because the second RCS firing added to the work that had to be completed in the last minutes of the count.

At the periscope where I could look over my right shoulder and see Jim Kelley, so after the recycle, everything went smoothly from there on until the count reached zero. The first-stage engines fired but then shut down! The ground launch sequencer detected that switchover of the launch vehicle's redundant guidance-flight control-hydraulic system

---

[1]   Private interview with James Kelley at Gemini fortieth anniversary, April 2006.

had occurred, which was a "NO-GO" for launch if it were to happen before liftoff. That was a mission rule—the vehicle couldn't fly if one of its redundant systems was not working correctly. So there we were with a fully loaded and armed launch vehicle still sitting on the pad, and I didn't know much about why it was still there! Bear in mind the launch vehicle test conductor knew more than I did because he was sitting by the ground launch sequencer operator and they had a display of sequencer lights, telling them what was wrong. He wasn't saying very much on the net except to get the various system guys to safe their system and that he wanted me, as the pad control, to get a recycle team to the pad ASAP. So I obediently went down to the ground floor of the blockhouse and donned the SCAPE suit. Along with about a half-dozen other people, we rode the truck out to the pad. The first thing I did as the pad control was get on the voice network to the blockhouse while the rest of the guys went around looking and installing locking pins in the destruct initiators. Talking on the net while in a SCAPE suit sounds like you are in a well, and it's hard to hear also. Everything around the first-stage engine compartment was dripping water from the flame bucket cooling water system having been turned on before the launch attempt. Also, all of the work stands from the first-stage engine area had been removed, and with that bulky SCAPE suit, it felt like there was danger of falling down the flame duct or tripping when straining to get a good look at the first-stage engine.

Jim asked me if I saw anything strange or unusual around the engine compartment, and I replied that there was a stream of fuel from one of the engine bells flowing down into the flame duct. Titan fuel was clear like water, but there was no mistaking what it was, considering its source and the steadiness of the stream. It was hard to hear in those SCAPE suits, and when he asked how big it was, I replied it looked about like the size of a pencil lead. Evidently, everyone in the blockhouse thought I said "big as a pencil," which shook up everybody listening especially the Aerojet reps (the engine manufacturer). I didn't see anything else unusual; and soon, just before the thirty minutes of air in the SCAPE suit had been exhausted, I was relieved at the pad by Ken Shipe, the second-shift pad control.

Robert L. Adcock

Subsequent investigation revealed a hydraulic actuator on stage 1 had burst open when the engine started, dumping out the hydraulic fluid; and the malfunction detection system, sensing this incorrect condition, commanded switchover to the redundant system because it knew it couldn't fly without hydraulics (there was a backup hydraulic system on stage 1). When this switchover occurred, the ground launch sequencer said, "Uh-oh," I see a "NO-GO" and promptly issued a shutdown command to the engine!

I didn't find out about this hydraulic problem until I stopped by the blockhouse on the way back to the office. I just hadn't looked up high enough in the engine compartment to see the damaged actuator. The stage 1 engine had undergone a hard start, and the kick caused so much backpressure on the hydraulic system that one of the actuators burst.

The remedy was to recast the actuator aluminum housings more robustly so they would be able to withstand future overpressures like this, if they happened again. Moog Servocontrols Inc., who built the actuators, worked with Martin engineers over the Christmas and New Year holidays to redesign and rebuild the actuators. The new, more robust, actuators arrived at the Cape on January 6, 1965.

The launch countdown began for the second time on January 19, 1965, and liftoff occurred at 9:04 AM. The rocket carried the spacecraft 100 miles up and 2,100 miles down range before it plunged into the south Atlantic some eighteen minutes later. The spacecraft heat shield endured this very severe reentry test. Testing the heat shield was one of the major objectives of this flight.

Chapter 6

First Manned Flight

Getting ready for the first manned flight had its moments, but they were pretty quiet compared to the launch vehicle first flights. Most of the bugs had been ironed out of the launch vehicle by this time, and the checkout went as planned. Not so for the spacecraft since it was the first full-up-configuration, manned spacecraft to be flown. I am sure McDonnell and NASA had their problems. I guess we on the launch vehicle were under their schedule umbrella most of that time.

In April 1964, the US Air Force had renegotiated the contract with Martin. The new contract went from a cost-plus-type contract to a cost-plus-incentive-fee type. The original contract was one of total cost reimbursement and a fixed fee, the fee being Martin's profit. The cost-plus-incentive-fee contract made the percent of profit a function of how well we did in delivery, launch, and performance of the launch vehicle. We, Martin at the Cape, had a large part of this. Martin could earn more profit on the incentive contract than on the fixed-fee type because of the increased risk. The new incentive contract also required only fourteen launch vehicles, whereas there were fifteen vehicles required at the outset of the program.

The entire contract rode on performance. We had to perform to meet the schedule incentives, and mainly we had to perform as a hardware provider and as an organization. The hardware had to deliver the spacecraft safely and on time to where it wanted to go, and the launch organization had to perform to get the launch off when the customers wanted it. The Air Force had put the incentive money on the things that were important to the program. The amount of the award was to be presented at the end of the program and was to be based on an objective grading system of our accomplishments. This system of observing the accomplishments and the ground rules that determined whether or not we got the incentive was worked out between Martin and the Air Force through negotiations. Martin had a contractual person in Baltimore who was the incentive monitor. His job was to keep all these ground rules in view and make sure the observations made by the customer were as essentially observed by Martin.

The launch vehicle team at the Florida launch site played a prominent role in determining our award fee. Therefore someone had to be assigned to monitor all the activities associated with launching and report on these soon after they happened. I was assigned to be the Martin Company Canaveral Operations mission monitor. One of the first things we did was to write the ground rules that Canaveral ops would be measured against. This involved the incentive monitor from Baltimore visiting the Cape, sitting with the Air Force rep Colonel Carl Ausfahl, and discussing various scenarios. The colonel was a shrewd negotiator, and I learned so much while I sat listening to these two. There were two or three of these all-day meetings on at least two different occasions to determine the ground rules. One of the ground rules or scenarios, as an example, was whether Martin would be penalized if the astronauts were told by Houston to *abort* the flight even though the reason was not the fault of the launch vehicle. Other ground rules governed just which data source would be used to ascertain whether the launch vehicle had made a specific objective. There were lots of players in this flight game. Martin wanted to make sure that any errors were truly our own and we were not being penalized for someone else's mistake. With my mission monitor hat on, that was my job.

Gemini III lifted off from Cape Canaveral Launch Complex 19 on March 23, 1965, with astronauts Gus Grissom and John Young aboard. They had been chosen for this mission almost immediately after the flight of Gemini I. A standard policy for NASA was to name a backup crew for each mission should the primary crew, for whatever reason, become incapacitated. The backups for Gemini III were Wally Schirra and Tom Stafford. It was typical for the astronauts to give a popular name to their spacecraft. This one was named Molly Brown, from the musical show depicting the character Molly Brown who, in real life, was a survivor on the ill-fated steamship *Titanic*, which sank in 1912. It was a reference to Gus's Mercury mission during which the spacecraft sank in the ocean during rescue and almost cost him his life.[1]

The mission's time in orbit was to be limited to three revolutions around the earth. The objectives of this flight were to demonstrate and evaluate the capabilities of the spacecraft and launch vehicle systems and procedures for supporting future long-duration flights and rendezvous missions. The spacecraft objectives were to test the ability to maneuver the spacecraft in space using the Orbit Attitude and Maneuvering System (OAMS). First, the OAMS would be burned to separate the spacecraft from the launch vehicle stage 2, in orbit, which would add a velocity of approximately ten feet per second to the spacecraft. Next, the OAMS would be used to circularize the orbit (the initial orbit is always elliptical after arriving in space, and it takes energy to circularize it). Other maneuvers were designed to demonstrate the ability to do out-of-plane adjustments (sideways motion) to the flight path. Lastly, the OAMS would be used to reduce the velocity of the spacecraft and position its alignment in space for the upcoming retro-rocket burn that was the first required maneuver to slow down the spacecraft and aim it correctly for the recovery process.

Some concern in NASA for the possibility of the crew being stranded in space should the retro-rockets fail drove the need for some of these OAMS tests. They knew that if the retro-rockets were to fail, the OAMS could be used to position the spacecraft in an orbit from which it was sure to reenter, thereby negating the results of a retro-rocket failure.

---

[1]   Hacker, Barton C. and Grimwood, James M. *On the Shoulders of Titans*, NASA SP-4203, (US Government Printing Office, National Aeronautics and Space Administration, 1977), p.233.

Experiments were also carried along on this flight for the astronauts to perform, which was as usual for all the manned flights. One was highly significant in that during reentry, the plasma flow caused by the ablation of the spacecraft heat shield entering the atmosphere drowned out all electronic communication. This prevented any communication from ground to space and vice versa and was called the blackout period. The experiment was designed to inject fluid into the plasma, which would then lower the plasma's frequency enough to allow UHF signals to pass through, thereby overcoming the blackout period to some degree. This experiment was successful.

The splashdown in the Atlantic was short of the intended landing point. The aircraft carrier *Intrepid* was approximately sixty-nine miles away from picking up the spacecraft. Faced with this problem, Gus then called instead for a helicopter to pick up the crew and the spacecraft. Debriefings began when the crew was onboard the helicopter and continued aboard the ship for several hours. One interesting note was that a corned beef sandwich from Wolfies' Restaurant in Cocoa Beach was carried along on the flight. Mission planners provided the astronauts' regulation food, but due to the shortness of the flight, it was just there for the tasting and evaluation. The astronauts largely avoided it in lieu of the sandwich treat.[2]

---

[2]   *Ibid.* pp.233-237.

Chapter 7

Eight-Day Mission

Sometimes when I was not the test conductor on a launch, I would go to Houston Mission Control and represent the launch vehicle to the NASA flight controllers. I could also do my mission monitor job from there. It was a good place to observe our performance because data from all sites were reported there. Chris Kraft and his men always treated us special when we came out. These were fine guys, and they loved to play when not in an operation. Martin Baltimore always sent two guys from its design team there also. They were called the slow-malfunction guys. It was possible for the launch vehicle to have a very slow malfunction in flight without triggering any of the automatic systems that warned the astronauts. Some scenarios envisioned such a slow-acting malfunction that none of the flight safety systems would detect. Therefore only the people looking at telemetry records in real-time flight could detect this trend and advise the Houston operations people on what to do.

I went to Houston for the flight of Gemini V. Jim Kelley drew the test conductor assignment for that mission. My job was to be in the control room at the Manned Spacecraft Center in Houston during the launch countdown and powered flight. There wasn't much to do but stand by. We had been given a console and a communications headset. We had access to more data than the people in the blockhouse control room at the Cape who were getting ready to fly the vehicle! We could talk to the people in the blockhouse or to anyone on the net there or at Houston. Our primary interface was with the Martin test conductor running the show in Florida.

When I arrived in Houston, I rented a car and drove out to the Manned Space Craft Center (MSC) and began to get my access badge to the Gemini control room. Chris Kraft had a "girl Friday" that looked after the flight controllers and visitors and made sure they knew what was going on and where they should be and how NASA could find us in the off-hours should a problem arise. She was good at that job and very cordial. Her name might have been Sue, so that's what I'll call her. When I got there, I found out the launch schedule had, for some reason, slipped a day or so. There was no time to go back home to Florida and return before the countdown started, so the guys from Baltimore and I consulted with Sue, and we headed for a certain bar in downtown Clearlake. Soon we were joined by some of the flight controllers, and we had a nice party. The party ended late, but I was already home because the bar was in my hotel!

The next day, the countdown finally got under way, and we all pulled into work sleepy but up for the flight. At T-5 minutes, the test conductor polled all participants in the launch for a status before entering the final phases of the count. The mission monitor (me) had to give a "GO" for launch. When it was nearing the T-5-minute mark, I suddenly lost all communications with the Cape; and unless the test conductor heard my voice give a "GO", he would hold the count. It was sort of a panic situation, so I called the flight director on the net and reported the problem to him. Before he could reply, someone on the net said, "We got it, flight," and within seconds, my net connection was back good as new. In the meantime, using a regular black phone on my console, I had dialed the test conductor's console in Florida; and Jerry Walden, the countdown announcer, answered. I told him to just keep the line open until after launch. Nowadays, that seems antiquated; but in those days, it was extraordinary that you could just dial the phone number—a long-distance call with no prior authorization—and get through! NASA Operations in Houston had the best setup for communicating I had ever seen.

On that flight, the launch vehicle flew perfectly; and then ten minutes after liftoff, as planned, it was over with for me. But the spacecraft began having problems, and I got interested listening to the talk between flight, the astronauts, and the flight controllers about their problems with the fuel cell. This was the first time a fuel cell that was to provide the electrical power to the spacecraft had been flown. Always before, batteries had done this job, but now systems were getting more complex, and more electric power was demanded than batteries could provide. Now the fuel cell oxygen supply tank pressure began to fall, indicating some sort of malfunction, and the fuel cell was unable to supply all the power needed. This did not bode well, and it looked on the surface like the mission might have to be terminated early for lack of electrical power.

Chris Kraft, the NASA flight director, began to talk to the astronauts and the ground flight controllers about conserving power and turning off the systems that were currently not being used. He began asking his electrical people in Houston about how many amps each piece of equipment demanded and how long batteries would last if this or that were turned off.[1] I used to be the electrical engineer on the Vanguard program, and I'm sitting there wondering how one guy could answer all those questions. These were the days before personal desktop computers and slide rules were still the mainstay of the engineers. I wandered down the hall, and I looked in a door of a room that was open. There were dozens of guys in that room figuring out the answers to those questions! I was dumbfounded! They had the best setup there I had ever seen for controlling the flight. If you wanted to know where the spacecraft was at any instant in time, all you had to do was look at a world map in the main control room and the ground track on the spacecraft were displayed on the earth map.

The flight controllers worried about this power problem throughout the entire flight with astronauts Gordon Cooper and Charles Conrad who flew that August 21, 1965, morning. Cooper and Conrad had been selected primary astronauts for this flight in February 1965, and Neil Armstrong and Elliot See backed them up. The flight could not have

[1]   Hacker, Barton C. and Grimwood, James M. *On the Shoulders of Titans*, NASA SP-4203, (US Government Printing Office, National Aeronautics and Space Administration, 1977), p.257.

lasted eight days were it not for the fuel cells providing the power. It turned out that even with reduced tank pressure, the cells continued with adequate power generation to see them through the flight.

Conservation of power was the tact Kraft took. Only thirteen hours of battery life could be counted on if the fuel cells failed! The fuel cells stabilized and continued to operate at less-than-full power output. The mission was continued, but the inside of the spacecraft became so cold due to the scarcity of electric power that the crew stopped the flow of coolant in their suit loops so they could warm up.[2]

Despite this power problem, the radar was later turned on and checked by tracking a transponder located on the ground at Kennedy Space Center in Florida. Normally, the transponder would have been on the target vehicle to aid in rendezvous with the spacecraft, but no target had been planned for this flight. This was a good first test because the radar was flying even though the transponder was on the ground. Later, a rendezvous with a hypothetical Agena was simulated by Houston flight-control people by sending up coordinates of a mythical Agena target and allowing the spacecraft to fly to these coordinates, simulating catching up to the target.

The fuel cells, as a by-product of power generation, also produced drinking water. Now, however, they were overproducing; and too much water was being generated. The astronauts couldn't drink all the water being produced. Sometimes power was turned off the systems just to stop the water production!

Although the fuel cells continued to produce power at a reduced rate, by managing its use, the mission was able to complete its planned time in orbit. At nighttime, the electric power was turned off, and the spacecraft drifted through space uncontrolled. Drifting meant that the spacecraft was uncontrolled in pitch, yaw, or roll because the power to the control systems was turned off. On day 3 in orbit, they did a number of experiments, including the tracking of the mythical target. In spite of these problems, the crew did a credible job on all the experiments.[3]

---

[2]   *Ibid.* pp. 257-258.

[3]   *Ibid.* p.262.

Hurricane Betsy moved into the planned landing area, and although Gemini could land in a more volatile sea state than Mercury, it could not land in a hurricane! The weather bureau recommended that Gemini V be brought down early to avoid the storm, but instead NASA changed the recovery site. The recovery ship, *Lake Champlain*, was repositioned to the new splashdown area in the Atlantic. On the morning of August 29, they fired the spacecraft retro-rockets; and as they started reentry, the reentry gauge showed that they were higher up than they should be. And with that, which may also be another first, Conrad yawed the spacecraft to ninety degrees to its flight path, causing the most atmospheric drag, thereby getting rid of the excess altitude. They landed about eighty-nine nautical miles short of the recovery ship due to an error in the spacecraft computer.[4] They had accomplished most of the mission's goals even though plagued by the lack of electrical power, but they had lasted eight days in orbit, one of the main mission requirements!

---

[4]   *Ibid.*

GEMINI 7/6 RENDEZVOUS
Gemini VII spacecraft as photographed by Gemini VI crew,
December 15, 1965, 160 miles above earth.

Atlas/Agena Launch
10 a.m. (EST)

Gemini VIII Launch
11:41 am (EST)

GEMINI VIII/AGENA MISSION
March 16, 1966

Launch Pad 19 showing the Gemini Space Vehicle standing with
umbilicals connecting to ground equipment and the erector in the
stowed for launch position. Some launch Team

Author seated at Test Conductor's console
in blockhouse at comlex 19, Florida

GEMINI IX-A
Augmented Target Docking Adapter

Launch of Atlas/ATDA at 10 a.m. (EST), June 1, 1966,
Cape Kennedy, Florida

Gemini IX-A and ATDA are only 66½ ft. apart.
Below is Caribbean Coast of Venezuela.

ATDA as seen by Stafford and Cernan from Gemini IX-A
spacecraft only 38½ ft. away.

Gemini IX-A approaches within 29½ ft. of ATDA
during rendezvous in space.

Launch Pad 19 showing the Gemini Space Vehicle standing with umbilicals connecting to ground equipment and the erector in the stowed for launch position. Some launch team members are standing in front

Gemini Titan 3 First Manned Launch, April 8, 1964.

1966--Martin Company Gemini Launch Team--1966

Chapter 8

A One-Two Punch

The Gemini rendezvous missions required a target for the spacecraft to link up with in space. This target was used by the Gemini spacecraft for rendezvous and docking—one of the prime program objectives to be demonstrated. Rendezvous meant they got close enough to almost touch the target, but docking was the physical act that connected them together in a rigid body. The target was an Agena rocket upper stage, which was flown into space atop an Atlas rocket. Once the Atlas had done its job of boosting the Agena to nearly the correct orbit parameters, the Atlas separated from the Agena, and using its own engine controlled from Houston, it was positioned in the correct orbit to become the target for the Gemini spacecraft. NASA could control the Agena engine from the Mission Control Center in Houston, and the astronauts could control it from their spacecraft. On more than one mission, with the Gemini spacecraft docked with the Agena, the Agena engine was fired, propelling the combined Gemini and Agena unit into higher orbits—where men had never been before!

Getting the Agena to orbit was done the same day as the Gemini launch. The Atlas Agena was launched from launchpad 14, about two miles east of the Gemini pad 19, approximately ninety minutes ahead of the Gemini. On pad 19, we had to be inside our blockhouse when the Atlas Agena flew, and we couldn't be counting down at the same time. If the Atlas Agena flew on schedule, there was only a short time hold in our count—five minutes or less to adjust our liftoff time so that the Gemini would have the correct orbital insertion time. Normally, the target would make orbit and circle the earth a couple of times before the Gemini caught up to it in space and started their rendezvous maneuvers. But the Agena had to make at least one partial earth revolution before mission controllers could determine its actual orbital parameters. These were then used to adjust the Gemini liftoff time and the rocket flight direction so that the rendezvous in space was efficient and within the capability of the limited propellants onboard the Gemini spacecraft. In the early rendezvous flights, the meeting of the target and the Gemini spacecraft occurred after three earth revolutions. To the launch team on Gemini, this required the Gemini launch vehicle test conductor to keep up with what was happening at the Atlas launchpad and adjust the Gemini launch countdown to the correct liftoff time. The onboard guidance systems took care of the launch azimuth adjustments, and these systems had to be updated from the ground based on the actual target-orbit tracking data.

The original plan called for Gemini VI to be a rendezvous flight with its Agena target. The plan was when the spacecraft got to orbit, the crew would maneuver into an orbital plane where they would catch up to the target and dock with it. Gemini VI would be launched after the target vehicle. Then the spacecraft would catch up with the target in about three more earth revolutions and dock with it. This series of two launches was attempted on October 25, 1965, but the target vehicle failed to achieve orbit. The launch of Gemini VI was scrubbed since to continue without a target vehicle would have been pointless. The question arose: what to do with the Gemini VI mission now that the target vehicle had gone and there was now nothing for it to rendezvous with?

Mission planners then planned a, novel, dual mission, utilizing two Gemini spacecrafts—something that had not been done before on Gemini or any other program. It became known as the Gemini 76 mission. In their plan, Gemini VII would be launched first, and since it was configured for a fourteen-day stay on orbit, this long-term objective could be partially accomplished while waiting for Gemini VI to be launched within ten days. With two spacecraft orbiting the earth at the same time, Gemini VI could rendezvous with Gemini VII in orbit, but not dock as there were no provisions for docking on either spacecraft. As far as the launch vehicle was concerned, this meant we had to check out and launch two, within ten days, off the same launchpad! We called this the rapid turnaround concept. Time between checkout and launch had been averaging about two months, and now they wanted us to launch two within ten days! The first I heard of this, Markus Goodkind started going to meetings where these plans were being hatched, and I remember thinking they were crazy! But as we reflected on the plans, it began to sink in that maybe it might just work. I guess Markus was included in the discussions since he was the test conductor when the launch of Gemini VI was scrubbed. Anyway, there were these daily meetings, and Markus was keeping everyone in the launch operations group apprised of the plans.

According to Joe Verlander, Martin's Gemini Florida director, he suggested the rapid turnaround concept. In his oral history report taken by the US Space Walk of Fame, he reports that in a think-tank kind of session called by NASA's Chuck Matthews, one of the ideas that came down was why couldn't we launch from the same launchpad on subsequent days and therefore speed up the launchpad productivity? It was the germination of an idea that came to its fruition in the Gemini 76 mission. It was Joe who followed up with the idea after he got back to Florida and put some meat on these bones that would work so well for the program. At that time, Verlander said the entire project was over the cost and behind schedule, and Matthews was looking for ideas that would bring the project back—put some showmanship into the program![1]

The logistics of this plan were terrible. To implement it, Gemini VI, which had been ready to launch but was scrubbed, would have to be removed from the launchpad and put into secure storage somewhere. That was so the launch-ready condition of Gemini VI, the launch vehicle and spacecraft, could be preserved until needed.

[1]   Verlander, Joseph M., (US Space Walk of Fame Foundation, Oral History interview, 2003).

Instead, Gemini VII, Titan and spacecraft, would be erected on pad 19 for normal checkout and then would be launched about two months later in mid-December 1965. The plan specified that when Gemini VII was safely in orbit, then Gemini VI would be returned to the launchpad and launched within ten days.

From the start of the program, NASA never had planned more than one Gemini flight at a time. The ground equipment in Houston's Mission Control was limited to one flight in orbit at a time. Ideally, back at the launch site, we would have liked to have two launchpads. That way, we could have had a vehicle on each pad each undergoing checkout independently of the other and then launch them on subsequent days. But that was not to be. Launchpads were costly and not justified. So the launch site people were trying to come up with a solution to the problem of how to get two launches off within ten days. We would have to forgo extensive launchpad refurbishment, and if the spacecraft and launch vehicle remained in storage while maintaining their launch readiness, a ten-day turnaround was doable!

This would be the first time four American astronauts would be in orbit at the same time. Launching again within ten days was contingent on minimizing the amount of launchpad and ground equipment damage that was experienced from the first launch, thus minimizing the amount of time needed to return the launchpad to a state ready to accept the next vehicle. By this time in the program, we had launched enough vehicles to know where the launch damage problems would likely exist. So we started laying in spare hardware to replace those damaged items right after the first launch. We also had a subcontractor standing by to give us a hand with steel replacement and the big-ticket items that needed welding. Normally, postlaunch refurbishment required about thirty days to accomplish. The time between starting checkout and launching took another thirty to forty-five days. We had, at most, about two days to get Gemini VI back on the pad and another eight days to get it ready and launched!

Mission Operations in Houston figured out how to support two Gemini spacecrafts in orbit at the same time. They studied the problem and decided that while both were flying, one of the spacecrafts would be designated active, and the other would be passive. When only Gemini VII was in orbit, Mission Operations would handle that operation normally. But when Gemini VI was launched, attention would be focused on it, and VII (having been in orbit longer, and therefore probably more stable) would be assigned the passive mode. Its data would be recorded at each pass over a ground-tracking station and then teletyped to Houston as had been done during the Mercury program. Data from the passive spacecraft would be reviewed after the fact and not in real time as was the data for the active spacecraft. This was how the mission control people planned for two in orbit at the same time. With the time constraint between launches from one-pad hurdle had been overcome and technical constraints in the ground and flight operations equipment had worked out for support of two spacecraft in orbit simultaneously, the planned Gemini VII and VI, i.e., 76 mission was "GO".[2]

Mission VII was launched normally on December 4, 1965. The astronauts were Frank Borman and James Lovell. The backup crew for Gemini VII was Edward White and Mike Collins. Borman and Lovell had been chosen for this mission in July 1965 and had been training for it ever since. The first men to attempt rendezvous and docking in space were supposed to have been Wally Schirra and Tom Stafford. That was because they were picked to be the prime crew for Gemini VI, but the flip-flop in plans for the 76 mission meant that Borman and Lovell in Gemini VII would get a crack at it first.

Since it would be ten days before Gemini VI would arrive on orbit, the Gemini VII astronauts were to demonstrate parts of the long-duration flight objective while awaiting the launch of Schirra and Stafford. Unique to their flight, the astronauts would attempt to work without space suits in the *shirtsleeve* environment maintained by their spacecraft. New lightweight space suits had been developed and would be used when undergoing hazardous maneuvers such as launch or landing. But in normal in-orbit operations, one astronaut kept his space suit on while the other astronaut doffed his. The cabin

---

[2]   *Titans*, p.273.

environmental control system did not work as well with the suits on as it did when they were off, so the temperature was always too hot for the one who had his suit on. The suits were hard to put on and take off in the limited cabin space of the spacecraft. Therefore Borman kept his suit on most of the flight while Lovell, the physically larger of the two and the first to take his suit off, was in shirtsleeves for the majority of the flight. Later during this flight, NASA relaxed its position on keeping one astronaut in a space suit at all times, and they both enjoyed the *suitless* environment.[3]

After spacecraft insertion into orbit, the Gemini VII crew kept up with the second stage of the launch vehicle that had also reached orbital velocity. They were able to rendezvous with the tumbling second stage, coming within distances of sixty feet of it, and they kept this up for fifteen minutes. Flying in formation like this was dubbed station keeping. Several experiments, as well as housekeeping of the spacecraft, kept the crew busy during this interim time waiting for the launch of Gemini VI.

Meanwhile, back on the ground, we went through the abbreviated launchpad refurbishment and erected the Gemini VI launch vehicle and spacecraft back on the pad. Nothing had been tried like this before! We were anxious to get started testing again so that it would confirm our judgment that a launch-ready vehicle could maintain its readiness to launch even though it had not been serviced in any way (other than a passive inspection of gas pressures in the tanks). The same logic was supplied to the spacecraft, i.e., it could be maintained in a launch-ready condition even though its maintenance in storage was little more than superficial for two months.

The launch of *Gemini VI* was attempted on December 12, 1965. When the countdown reached T-0, the first-stage engine roared to life and then shut down, and there was no liftoff. Astronauts had both liftoff and engine-shutdown lights displayed in their spacecraft. These were all the ingredients needed for them to decide to abort, exiting the spacecraft by using the escape rockets attached on their seats.

---

[3] *Ibid.* pp.277-282.

Robert L. Adcock

Under the mission rules concocted before the launches, the presence of these two lights pointed to a condition where they were flying and the engines quit running, which was a prescription for disaster unless they bailed out. This was not the condition at launchpad 19. Engines had started then shut down, and a false liftoff indication was given, but there was no vehicle movement. As it was, cooler heads prevailed, and astronaut Schirra (the commander) elected to remain with the spacecraft and exit normally, which was when we could raise the erector and let them step out. Stafford followed Schirra's lead and stayed in the spacecraft. Schirra's earlier experience on the Atlas/Mercury launches probably aided him in his decision. He said that even though he had a liftoff light, he did not feel the booster move from the launchpad and therefore knew they couldn't have been flying, so no abort was attempted. Frank, the chief test conductor who was supposed to advise them during this time, was cool and reassured them of the situation. When troubleshooting why the launch vehicle shut down, it was found that one tail plug that was supposed to separate at liftoff had inadvertently separated at ignition and caused the ground equipment to shut down the engines.

Later in my life I had a chance to talk to Wally about this incident. He was a guest of NASA at Kennedy Space Center. Cal Fowler from the Space Walk of Fame and I had the chance to see and interview him between his public appearances that day. I broached the subject, saying that the astronauts had all the data they needed to initiate an abort from the spacecraft. They had the liftoff display and engine-shutdown lights, and yet they did not abort. He said, "Yeah, it just didn't feel right!"

During this launch attempt, I was supervising second shift (we worked two twelve-hour shifts until the launch) and was home in Cocoa Beach because the launch attempt was in midmorning. I got out of bed to watch it. I had been up all the previous night. I usually watched the ignition sequence on TV and then stepped outside the front door to watch the flight. I saw the engines light on TV and then walked outside. But where was it? I rushed back inside, and there it sat on the launchpad! Was I nuts? Didn't I just see the engines running? All the TV newsmen were talking about what happened. The way it worked was we would start the first-stage engines, then wait almost two seconds while thrust built up. If the engines were working OK, then the ground launch sequencer would fire four explosive bolts that held the vehicle to the launchpad, thereby releasing the vehicle for its flight. I had seen the engines start on TV, but it didn't release, and she was still standing on the pad! That meant the bolts hadn't blown—for any number of reasons. One of the newsmen reported, "Big arms reached up and grabbed the vehicle and kept it from flying." We all got a chuckle out of this! There were, of course, no mechanical arms that returned the vehicle to its resting place on the pad!

My first thought was about having to go back into work again that night because we were planning a short shift. In fact, I had planned to stay home. It was Sunday night going into Monday morning. Now the *Gemini VI* vehicle turnaround for launch had to be completed in time for Schirra and Stafford to meet Borman and Lovell, who were already in orbit. By this time, *Gemini VII* had been in orbit eight days, but it still had adequate fuel left to complete the fourteen-day mission.

Engineers looking closely at the telemetry data from *Launch Vehicle VI* later discovered that one of the first-stage engine subassemblies had not started to build up power as it should have, and that needed further troubleshooting. Subsequent inspection revealed a plastic dust cover had been inadvertently left in an oxidizer gas generator line on the engine when the engine was attached to the Titan stage 1 in Baltimore during vehicle assembly! All of the testing subsequent to that had not shown this error. Getting over this problem doubled the work for an already-tight turnaround and recycle schedule.

The first night shift after the shutdown was spent unloading the fuel and oxidizer from the tanks and taking out the ordnance. The spacecraft guys had to do the same sorts of things, and we had to work around one another. Later in the day on Monday, when I heard the news over the car radio that the plastic dust cap had been found in the engine, I couldn't believe it. On the following Tuesday, as I reported into work, Frank took me aside in the ready room and told me we had a lot of work to do to get ready for countdown pickup and launch on Thursday.

When Frank showed me the schedule board, I almost fainted. There looked to be a normal week's worth of work to do in one night! Most of the work was out at the pad on the first-stage engine. Once the engines had been exposed to propellants, as was the case when the engine fired but shut down, an inordinate amount of cleaning and restoration of certain gaskets and one-time-used items had to be replaced. This entailed not only the mechanical replacement but also a lot of leak checking after replacement of these items. After the day shift went home, I called George Taylor, the quality manager, and he and I went out to the pad at the base of the vehicle and set up a workplace. I then got in touch with all the propulsion engineers and techs. I told them I wanted them out on the stand, simultaneously running tests on both subassemblies of the engine. George had his quality inspectors there too. I told this team that if they needed parts or paper, someone would get it for them; and if they needed a smoke break, to take one there at the edge of the pad ramp. I asked the radio frequency guys (who weren't so busy) to make a forty-eight-cup pot of coffee and set it alongside the work area. When the range contractor chow truck came on the pad at midnight, I bought all the donuts she had, several dozen; and we had coffee and donuts for the rest of the evening. This was unheard of! There were safety regulations preventing smoking within five hundred feet of the launch vehicle. Drinking coffee at the launchpad was probably also prohibited in our rulebook. But everybody wanted to get the work done, and we bent the rules so there was little interruption to the test work. By morning, we had all the engine work done.

Range contractor and NASA support to the launch operations people were being provided around the clock. Ordinarily, range contractor support was almost nil in the wee hours of the morning unless it had been prearranged, but they were there to help any way they could during this ten-day turnaround.

At the end of that shift, I went over to the ready room where the multitudes were gathering for the seven o'clock morning meeting, and I whispered to Frank that our work was complete and we were on schedule to pick up the count for a launch on Thursday morning. This news made him mighty happy! I think in his mind, he really had a serious doubt that the work could be done in such a short time frame.

So NASA and the Air Force came to the seven o'clock morning meeting, expecting a lot of work yet to be done, probably; but when they heard we had completed this big chunk of work the night before, they were elated. NASA had all its helpers and VIPs there: Dr. Mueller, NASA deputy associate administrator for space flight, and Charles Matthews, NASA Gemini program director. Newsmen from the TV networks were also there. We went ahead and picked up the count; and on December 15, Wally Schirra and Tom Stafford, aboard *Gemini VI*, were successfully launched into orbit.

Immediately, the crew of *Gemini VI* began the rendezvous procedures. Schirra and Stafford were flying the chase mission, and their radar locked on when they were 434 kilometers (270 miles) apart.

Schirra and Stafford continued rendezvous procedures, adjusting orbital plane, adding velocity and altitude until the two crews could see each other forty meters apart. They station kept with each other for three and one-half orbits. The crew of *Gemini VI* flew around *Gemini VII*; and when they were in view of *Gemini VI*, the crew of *VII*, Schirra and Stafford, held up a hand-printed sign in their window that read Beat Army, reflecting the recent army-navy football game. Schirra was a graduate of the Naval Academy, while Frank Borman was a West Point alumnus. Incidentally, the outcome of that game was a 7-7 tie and starred navy quarterback Roger Staubach who went on to have a brilliant career with the Dallas Cowboys.

NASA awarded the Group Achievement Award to the Gemini VII/VI launch operations team *for exceptional achievement by NASA-Air Force-Industry launch support crews, which in eleven days successfully prepared and launched* Gemini VI, *making possible rendezvous with* Gemini VII.

So many of the program objectives were accomplished during these two flights—especially the ability to maneuver the spacecraft, change planes, and station keep such that rendezvous would not be a problem. Following rendezvous, the Gemini VI came back to earth on December 16. Gemini VII demonstrated the fourteen-day mission, and it returned to earth on December 18. Bob Gilruth remarked that "we have accomplished the major part of the Gemini space objectives at this point."[4] Lucky for us he didn't stop the program there although with all the Apollo spending, some thought it might have been justified.

---

[4]   *Ibid.* p. 295.

Chapter 9

First Man Makes First Rendezvous

A prime objective of the Gemini VIII flight was to rendezvous and dock with the target vehicle, one of the prime program objectives. This had been thwarted before when the target vehicle failed to make orbit. The act of physically docking to another vehicle still had to be demonstrated, although the Gemini 76 mission had worked out the rendezvous procedures. The Apollo moon program needed to get this achieved because they were planning to rendezvous the Command/Service Module with the Lunar Excursion Module on the way to the moon and then dock the Lunar Escape Module to the Command Module when returning from the moon to earth. The Agena's failure to support Gemini VI's flight caused the whole target fleet to be grounded while a failure analysis team, consisting of NASA, Air Force, and contractors, determined the Agena's problem and reworked it. The team had a short time, and they would take over four months to find and engineer a design mod for it. The problem was when the engine was started, fuel entered the combustion chamber before the oxidizer, as Agena was designed; and that sometimes resulted in a "hard start," which is defined as unstable burning in the engine thrust chamber. The team called a symposium at the Kennedy Space Center to explore the cause, and nobody in the seminar could come up with anything that could explain the problem any better than the "hard-start" hypothesis. The repair for this problem consisted of converting the starting sequence in such a way that the oxidizer always led first into the combustion chamber before the fuel. Although the correction sounded simple, it was far from it. It meant modifying and requalifying the engine. Several engine starts at simulated altitude were required to requalify the engine. The only place these tests could be done was at the US Air Force's Arnold Engineering Development Center in Tennessee. The Air Force gave the Agena testing priority over their other work to support the Gemini mission. Around-the-clock testing of the engine, with the good results they experienced, enabled the engine and target vehicle to be requalified barely in time to support the flight of Gemini VIII.[1]

Another task for this Gemini VIII flight was to test extravehicle activity more than had been done on *vehicle IV*. The USAF had developed an Astronaut Maneuvering Unit (AMU) to be used to propel the astronaut outside the spacecraft. It was too big to fit inside the cabin and was bolted outside to the spacecraft in the aft equipment area. To don the AMU, the astronaut would have to get out of the spacecraft and maneuver himself to the rear adapter section before he could don the unit. The adapter section was the compartment between the launch vehicle second stage and the spacecraft cabin. All the while he was supported by oxygen and environmental-control umbilicals from the spacecraft. Umbilicals were known to be bad for coiling uncontrollably during extravehicular activity, as Ed White found out on Gemini IV.

With all the Agena engine mods completed, the next target, the one assigned to the Gemini VIII mission, was launched from pad 14 atop an Atlas rocket on the morning of March 16, 1966, ninety minutes ahead of the planned Gemini VIII launch. The target vehicle launch time and orbital parameters were crucial in determining the liftoff time of the Gemini itself. Early tracking data from the Atlas Agena launch was factored into the Gemini vehicle's ground guidance system, and a liftoff time and the roll program necessary for that rendezvous were computed. The launch vehicle liftoff time (manipulated by the built-

---

[1]   Hacker, Barton C. and Grimwood, James M. *On the Shoulders of Titans*, NASA SP-4203, (US Government Printing Office, National Aeronautics and Space Administration, 1977), pp.298-303.

in hold time in the countdown) and roll program (set in by ground guidance) were adjusted to accomplish these orbital outcomes. This resulted in a desired liftoff time of 10:40:59 AM. On these rendezvous launches, both the Atlas Agena on launchpad 14 and the Titan Gemini on launchpad 19 started their launch countdowns about the same time. Periodically, personnel from each launch operation would check with one another to see how it was going. The Agena people didn't want to launch unless there was a high probability of a Gemini launch. After the Agena was launched, the Gemini count went into a hold; and after the desired liftoff time of the Titan Gemini was known, the resumption of the countdown time was calculated by the test conductor to meet the required liftoff. On this mission this liftoff time was perfectly achieved.

I was test conductor on this launch. A few days before launch, Frank and I went over to KSC and met with the astronauts. The Martin people didn't see much of the astronauts during normal checkout and launch preps because they dealt mainly with the spacecraft people, McDonnell and NASA. The reason we went over to meet them was that the chief test conductor had responsibility for the astronauts' safety during the first ten seconds of the flight or until the vehicle cleared the umbilical tower, whichever happened first. The logic being that if an emergency situation arose during that short time, the astronauts would recognize our voices on the network as we were issuing commands and mentally view that face with that voice. We also reviewed with them the emergency procedures we had to execute on the vehicle if a contingency situation arose. The meeting never lasted longer than one hour. As I recall, only the prime crew was at that meeting. It was low-key, and the astronauts were pleasant and very humble.

Astronauts Neil Armstrong, a civilian test pilot who later would become the first man to set foot on the moon, served as the command pilot. David Scott, who was new to the astronaut corps at that time, was selected to be his flight mate. The backup crew was Charles Conrad and Richard Gordon Jr., both navy lieutenant commanders. Whereas Gemini VI and VII had rendezvoused in flight, the important objective of accomplishing a rendezvous and docking with an unmanned target had yet to be demonstrated.

Robert L. Adcock

After a near-perfect powered flight to orbit, the crew settled down for the first phase of the rendezvous mission, which was catching up in space to the target vehicle. They were about 1,963 kilometers (1,200 miles) behind it. They caught up to the target by selectively firing their thrusters to adjust their orbital speed to the target's speed. Other firings would adjust their orbital plane to match the orbital plane of the target vehicle. When they were about 205 miles from the target, the rendezvous radar locked on between the spacecraft and the target. When they were approximately 332 kilometers away (206 miles), the radar was switched to rendezvous mode, and the spacecraft was then piloted to within a few meters of the target. After visually inspecting the target, Armstrong eased the Gemini spacecraft nose into the target's docking mechanism, and full docking was then effected. When docked together, the two spacecraft formed a rigid unit that could be undone only by reversing the docking procedure and using power from the spacecraft thrusters to back away.

When the spacecraft-target combination vehicle was commanded to turn right by ninety degrees, it was noticed that the vehicle combination began to roll. It seemed the Agena target vehicle was causing the problem. They had been warned by Mission Control earlier that the target's attitude control system might give trouble based on what they, the ground crew, had seen when they were verifying the Agena's stored commands prior to docking. With this warning in mind, the astronauts immediately suspected the Agena. They tried to stop the rolling with the spacecraft OAMS but could not. They tried turning off the Agena's attitude control system. When they looked at how much OAMS propellant they had used trying to stop the roll, they began to think that maybe the spacecraft OAMS might be part of the problem. The roll rate got up one revolution per second, causing the astronauts' vision to get blurry. It was obvious that they wanted to separate from the target vehicle to eliminate it as the problem. When they separated from the target, their roll rate increased! That meant something on the spacecraft itself was causing the unwanted roll maneuvers. Only when they had activated the reentry control system and shut down the OAMS were they able to stabilize the spacecraft and stop the roll. It became obvious that an OAMS thruster stuck open and had been firing all the time. This episode happened when they were out of communication with the flight controllers in Houston because they were on the wrong side of the earth for the signals to get through.

Because the reentry control system (RCS) had been used during this emergency, mission rules required them to terminate the mission ASAP. The RCS was absolutely needed for alignment of the spacecraft for reentry, both before and after the retro-rockets were fired. Its fuel margin did not provide for in-orbit attitude adjustment. As soon as practical, the spacecraft was brought home. Because it was an unplanned reentry, it splashed down in the Pacific Ocean instead of the Atlantic where it was supposed to return. The destroyer, the USS *Leonard Mason,* was pressed into rescue service. They raced at flank speed and found the spacecraft after splashdown at a point in the Pacific, 800 kilometers (500 miles) east of Okinawa, Japan, and 1,000 kilometers (621 miles) south of Yokosuka. The mission lasted just ten hours and forty-one minutes.

Afterward, the flight controllers in Houston flew the target vehicle alone, performing several maneuvers sufficient to determine that the fault that caused the early mission termination was a "stuck open" thruster of the spacecraft OAMS, not the Agena. This target vehicle would be used again on a later flight.[2]

By the time I got home from the postlaunch party, my wife told me there was a problem, and they were already back on earth! I almost went into panic mode—I knew so little about the problem. I thought it might have been something the launch vehicle did to the Gemini spacecraft that terminated the mission. What relief when, going to work the next morning, I heard on the radio that NASA acknowledged the fault was in the spacecraft.

---

[2]   *Ibid.* pp.308-321.

Chapter 10

The Angry Alligator

Elliot See and Charles Bassett were assigned the command and auxiliary pilots respectively for Gemini IX. Their backups were Tom Stafford and Gene Cernan. Tom Stafford had already flown on Gemini VII. In February 1966, as part of the astronauts' training, they made a trip to St. Louis to visit the McDonnell factory where the spacecraft for Gemini IX was being assembled. The astronauts flew in two T-38 jets. Elliot See piloted one and Basset was his passenger while Stafford piloted the other. The weather there was not good, and flying had to be on instruments. When the two planes flying in formation broke through the overcast at Lambert Field in St. Louis, they were too far down the runway to land, and they elected to go around. See banked his plane left, but recognizing his sink rate was too great, he turned on the afterburner and then realized he was heading for the McDonnell assembly building and then, too late, turned right sharply. The flight momentum carried the plane into the roof of the building where they then crashed into its courtyard, killing them both. The other airplane with Cernan and Stafford was able to execute a missed-approach-procedure turnaround and land safely.[1]

---

[1]   Hacker, Barton C. and Grimwood, James M. *On the Shoulders of Titans*, NASA SP-4203, (US Government Printing Office, National Aeronautics and Space Administration, 1977), pp.323-325.

The backup crew, Stafford and Cernan, then became the prime crew for Gemini IX; and James Lovell and Edwin Aldrin were then named backups.

On Gemini IX, the use of the USAF-developed Astronaut Maneuvering Unit was a high priority. It was developed, initially, for their Dyna-Soar program, and following its cancellation, the AMU was furnished to NASA for use in Gemini. The AMU had flown before, but because of problems and the short flight of Gemini VIII, it was not donned. A debate ensued in NASA over whether an astronaut with an AMU strapped on and outside the spacecraft in orbit should be tethered to the spacecraft or could he be entirely free from all ties to the craft (the AMU providing all necessary life support and propulsion to the astronaut). NASA eventually ruled that all astronauts engaged in extravehicle activities (EVA) would be tethered on all remaining Gemini flights, whether or not with AMU.

Discussion in NASA also arose over whether radar was required to assist with rendezvous or the astronauts could carry out the task using only visual contact with the target vehicle. By eliminating the rendezvous radar, a significant weight saving for the Apollo program, currently in design, could be realized because the radar could be left off the Command Module and the radar transponder off the Lunar Module. NASA ruled that the radar would fly on Gemini, and experiments would be conducted designed to try spotting the target vehicle by eye when they were over the Sahara Desert, which was similar to the moon's surface color.

Additionally, on flight IX, a rendezvous with the target vehicle would be attempted as early as the third revolution of the spacecraft following its orbital insertion. Others had been later in the flight, but an attempt on the third revolution would simulate the Apollo's Command Module rendezvous with the Lunar Module in that flight plan, and therefore it became a goal of Gemini IX.[2]

The first attempt to fly Gemini IX was made on May 17, 1966. The test conductor for this launch was Ken Shipe. Shipe had served as both propulsion lead engineer and, later on, as assistant test conductor

---

[2]   *Ibid.*, pp.325-328.

during several launches. Now this would be the first time he had actually launched one.

The target vehicle failed to achieve orbit, and the Gemini mission was scrubbed. NASA had a backup plan for that case, which was to launch Augmented Target Docking Adapter (ATDA). The ATDA was a system that could be carried aloft by the same type of Atlas launch vehicle that would have put the Agena target in orbit. The ATDA would be in orbit and provide the target for the spacecraft demonstration, albeit a dead one (lacking guidance or propulsion). NASA had preplanned for this eventuality and could be ready to launch the ATDA within fourteen days.

The next attempt to fly was on June 1, but this time with the ATDA as target. The ATDA target launched exactly on time at 10:00 AM and then went into orbit. But it was found that the aerodynamic cover (the nose fairing) over the docking port on the ATDA had failed to jettison and was only partially opened. Even though the Gemini proceeded on toward launch, later a failure in the ground control system that updated the spacecraft with refined launch azimuth information caused the launch to scrub that day.

On June 3, Gemini IX launched at 8:39 AM, and the vehicle proceeded to orbit without incident. The Gemini caught up with the ATDA some four hours after launch and began station keeping with it. While station keeping and looking at their target from as close as thirty feet astronaut Stafford proclaimed that the ATDA, with the fairing that had failed to jettison, "Looks like an angry alligator out here rotating around." After several discussions about what might be done by the flight crew to release the fairing, all suggestions were soon abandoned as too risky for the crew and mission. It was impossible to dock with the ATDA with its fairing only partially jettisoned. As a result, they gave up on the docking objective. However, the crew accomplished several attempts at rendezvous with the crippled target in space. During these rendezvous attempts, the crew tried their ability to spot the target vehicle and concluded that without the rendezvous radar, this task would have been impossible.

The next day, it was time to try out the AMU. Cernan went outside the vehicle toward the rear to don the unit. It became immediately apparent that zero-gravity training simulations on earth had failed to provide him with knowledge of the difficulties he encountered in orbit under the actual zero-G condition. Almost any leverage he used to position himself in this environment caused him to move about wildly, and there were inadequate hand—and footholds with which to restrain himself. Nevertheless, he moved to the rear of the spacecraft in preparation for donning the unit. In the process, he became so hot and exhausted his face shield began to fog up and he could not see. He rested as much as he could until the fogging cleared and then resumed putting on the unit. The exertion required donning the AMU, plus the extra time he would have required to accomplish maneuvers, would have seriously depleted his energy, and the crewmen decided to scrub the AMU experiment.[3]

The remainder of the time in orbit was utilized doing various experiments. On June 6, some seventy-two hours after launch, the spacecraft landed in the Atlantic and was retrieved by the carrier USS *Wasp*. The location of the spacecraft relative to the carrier was close enough that the astronauts remained inside the spacecraft as it was being hoisted aboard the ship.

---

[3]   *Ibid.,* pp. 336-341.

Robert L. Adcock

Chapter 11

The Switch Engine

The target vehicle for Gemini VIII, the Agena, had been left in orbit after completing the mission. NASA flight controllers flew the target after the return of Gemini VIII to earth. It turned out that Gemini VIII's plight had been the stuck-open spacecraft OAMS thruster, and the target was perfectly good. NASA parked the Agena in a high orbit (approximately 250 miles circular) waiting for the next activity for which it could be used. Despite its loss of electrical power due to battery depletion, the North American Air Defense Command had been tracking this target and had a good understanding of its orbital parameters and where it was located.

Gemini X also had a target vehicle to be launched. If all went well, the Gemini X target vehicle would be launched, followed by the Gemini X spacecraft, carrying veteran astronaut John Young and Michael Collins. Their backup crew was Edwin Aldrin and Jim Lovell. They would link up with the Agena X target; and using its propulsion system and guidance, they would then proceed through maneuvers to move up to the orbit of Agena VIII target, rendezvous, and station keep with it. The Gemini spacecraft had insufficient fuel for so large a maneuver on its own and had to rely on the energy contained in the Agena to change orbital paths. This would be the first time a linked-up Gemini spacecraft ever flew with an Agena vehicle providing the power to change orbits. Station keeping with Agena VIII would be tricky since the batteries had gone down because it was left in orbit with no way to recharge them. There was no rendezvous radar on it. The astronauts had to depend on their eyesight to find it and perform maneuvers to get in the same orbital plane and get within a few feet of it. When performing all these scheduled maneuvers using Agena power, the Agena was dubbed the switch engine.[1]

The prelaunch preparation of the Titan went well, and Don Striby was assigned as test conductor for this launch. Don had risen from instrumentation lead engineer at the beginning of the program through assistant test conductor until this day when he would actually fly one himself. By this time, Chuck Cicchetti had left the program after the launch of Gemini III to go to work for IBM, and Jim Kelley left and went to work for NASA KSC on the Apollo program after the Gemini 76 mission. These were both senior people so their leaving made room for advancement within the ranks of the test conductor organization. On launch day, I actually watched from Launch Complex 16, another Titan launchpad, through the periscope as Gemini rose from Complex 19. This was standard procedure to get all the eyes on the "bird" you could, just in case we might see something if anything unusual happened during liftoff. It was easier to get to Complex 19 from Complex 16 if they required somebody to go there after they had closed the blockhouse door.

At about this time in the program, people started jumping off the program to go to other contractors and NASA "over 'cross the river" because they were building up for the Apollo program. People knew that Martin would have slim job pickings at the end of Gemini, and the time was ripe for getting good jobs on Apollo. In the aerospace business, this was a major personnel concern, that is, how to stay employed when programs came to an end and how to advance in position and pay. Martin began to be concerned that their launch team personnel would jump ship before the program was over and jeopardize Martin's performance. So on a Sunday morning, we were

---

[1] Hacker, Barton C. and Grimwood, James M. *On the Shoulders of Titans*, NASA SP-4203, (US Government Printing Office, National Aeronautics and Space Administration, 1977), p.342.

Robert L. Adcock

called to participate in a meeting at the launch complex called by Martin vice president Bill Bergen of Baltimore in which he announced a bonus to be given to the individual launch team members at the end of Gemini. This bonus was to be $200 for each successful Gemini flight in which the individual employee had participated. We had to stay to the end of the program to get the bonus, and we wanted to stay because the promise was for $2,400 if the whole launch program was successful! That sure slowed the outflow of key personnel from the launch team.

That was a lot of money in those days. I got paid the entire amount at the end of the program because all the flights were successful. I put the money in a savings and loan account and then went car shopping. I bought a new 1966 Mercury Monterey and paid cash, trading in my used 1960 Monterey I had bought in California. I blew it all in one afternoon!

Gemini X launched on July 18, 1966, and both the Agena and the Gemini flights were without incident. Five hours and fifty-two minutes after the launch, they docked to the Agena X target. While docked, they started the Agena's engines for propulsion. That caused the Gemini spacecraft to be pushed backward from its nominal flight direction, resulting in the acceleration forces felt by the astronauts as "pulling from the front." This was opposite the forces they felt from the back when they were being launched. It was a new experience for the astronauts. The first burn of the Agena added 288 miles per hour to their velocity, resulting in an oval orbit 475 miles (763 kilometers) by 183 miles (294 kilometers). This was the highest above earth that any humans had ever been before!

Now it was time for the Agena to maneuver the combination spacecraft-target vehicle through space to a position enabling a rendezvous with Agena VIII. After a crew rest period, the Agena fired to reduce their velocity by 243 miles per hour (391 kilometers per hour) and the highest point in their orbit to 237 miles (381 kilometers). An adjustment burn later added 56 miles per hour to circularize this orbit in preparation for finding the Agena VIII. The Gemini X Agena target vehicle obscured the field of view of the astronauts while it was docked with the spacecraft, and they could not spot the Agena VIII.

The next day, with the duty of the "switch engine" function complete, the Agena X was undocked. They then spotted the Agena VIII when it was approximately 20 miles (33 kilometers) away and rendezvoused with it. When they had steered to within 10 feet of it, Collins got out of the spacecraft and went over to the spent Agena VIII, got the micrometeoroid collection experiment package (that had been there since the Agena VIII flight) and returned to the spacecraft.[2]

The crew of Gemini X came back to earth on July 21, splashing down in the Atlantic, where the navy ship *Guadalcanal* recovered them.

---

[2]   *Ibid.* pp.344-351.

Robert L. Adcock

Chapter 12

Ride 'Em, Cowboy

Things on the launch vehicle side were going so well they were getting kind of ho hum. The vehicle was responding to its tests nicely, and there was not too much rework or modifications that had to be done. In the test conductors' office, we were thinking about the rest of our lives after Gemini. Disney World, near Orlando, was just getting going and attracting many tourists. While we were not thinking about going over and being part of the show itself, we would have been open to a job that would have let us use our new experiences and also given us a healthy raise. By this time we were starting to believe our own newspaper stories. Heck, we were told daily just about how smart we were and how the program couldn't get along without us!

Someone got the bright idea that we could own the campgrounds of America (KOA) Disney franchise, and that would be like gold! KOA already had name recognition and brand loyalty among the people who camped. Some salesman named Bob, not an aerospace person, had come along and knew about camping. Five of us anted $500 each, formed a small combine, and bought the franchise. With franchise in hand, we went looking for land close to Disney World we could lease, or otherwise acquire, for the campground. Others were busy trying to decide such things as the number of campsites an acre would accommodate and how many tent sites versus how many trailer sites to put in, how much revenue to expect, and most of all the investment necessary. And of course, the profit! We had a monthly progress meeting with the partners, and we documented all our trade studies like the good engineers we were. I was again working second shift when the last monthly meeting occurred. I hadn't gone in for it (the meeting was at the Cape Royal Office building in Cocoa Beach), but I got a call from one of the guys who had. It seemed that Bob, the salesman, got up in the meeting and said we were taking too much time and wouldn't agree to anything. Meanwhile time was wasting, and he had acquired a spot of land south of St. Augustine on the beach. He wanted to dissolve our little combine, buy the franchise, and get out of business with the other four of us. I don't remember his offer at the time, but I considered it about one day and decided to sell my share. The other three caved in too and sold theirs. Bob was true to his word and built a nice campground right off US 1 and invested $35,000 before he got ready to open. I am sure it was a success. So much for the idea of being a campground baron!

For Gemini XI, Charles Conrad and Richard Gordon were chosen as crew while Neil Armstrong and William Anders were named alternates. The objectives of this flight were ponderous, with major demonstrations and experiments planned. The program was nearing the end, and some experiments and demos had not been accomplished, so this created a full load for the astronauts' time in flight. Program management had established the last day of January 1967 as the date for the program's end. They wouldn't go beyond this point, and we knew it. *Vehicle XI* was the next-to-last vehicle to fly, and the manufacturing line in Baltimore had already shut down, pending completion of their *vehicle XII* launch support.

An idea was kicked around in NASA planning: flying a Gemini spacecraft around the moon, using the Agena target vehicle for propulsive power. To do this, Gemini XI would have flown as normal, docked with the target, and used the Agena engine to propel them into an orbit which would have included the moon. This idea was ruled out, but the seeds for a high-flying vehicle had been sown. Charles "Pete" Conrad championed the idea of a high flight and went about selling it to NASA management. One justification he used was that the weather bureau was considering colorizing the photographs from their weather satellite that was being developed. Conrad proposed he

Robert L. Adcock

take color pictures from this mission's planned high altitude and bring them back to help the weather bureau people in making their decision. He also got other experimenters to verify which experiments would be aided by the higher flight and got their concurrences.

Another experiment that was assigned to Gemini XI was one in which artificial gravity would be instigated by maneuvers of the spacecraft and its target. In space, it was thought man could endure weightless conditions for short periods only without deleterious effects to his health. Over a long time in orbit, however, some provision had to be made to overcome this gravity deficiency. This experiment would show the feasibility of generating an artificial gravity field by two different means while in orbit. This consisted of tethering the target vehicle to the spacecraft and separating the two by about one hundred feet, so that the tether between them would be taut, then rotating the tethered pair.

Two different configurations were to be attempted to generate this artificial gravity field. In one, called the gravity gradient experiment, the tethered-and-docked combination would be aligned perpendicular to the earth with the Agena engine nozzle pointing downward. After the vertical alignment was correct, the spacecraft then would undock and back slowly away from the target until it was one hundred feet above the target, with the tether taut. Separated by one hundred feet, the force of gravity on the Agena would be slightly more than on the spacecraft, and this difference in force (although miniscule) would cause the two objects to remain in the vertical configuration (theoretically). In this configuration, the vertical tethered assembly would continue in orbit, and the tether would remain taut due to the gravity gradient between the two.

The other configuration would have this same combination, still tethered, but spinning about a hypothetical pivot located somewhere toward the center of the tether at the rate of ten degrees per second. Centrifugal force generated by this spinning would not only keep the tether taut between them but also create artificial gravity. In this configuration, the two spinning vehicles would encircle the earth, joined together by the tether.

In another objective, the Apollo program suggested that rendezvous be demonstrated on the first orbital pass of the mission (called M=1 for the first revolution of the orbit rendezvous) to more closely simulate conditions of the Command Module and Lunar Module rendezvous for the moon flight. These maneuvers had been previously carried out during revolution four (M=4), which gave the spacecraft more time to catch up with the target vehicle. The effect of this early rendezvous on ground operations was to significantly reduce the launch window of the Gemini. The launch window is defined as that acceptable period in which the launch can occur. The launch window was short so that the distance between the target and the spacecraft would be short when the spacecraft arrived on orbit. This distance had to be overcome by burning the spacecraft thrusters and using fuel from its limited supply. Making the rendezvous in M=1 caused us to have to launch very soon after the target passed overhead, on its first revolution, at the very opening of the launch window.[1]

This was my flight as test conductor. Things were getting easier; on the night before the first launch attempt of GT-11, the second-shift crew set a new time record for loading the launch vehicle propellants in two hours, fifty-eight minutes, and nine seconds! This was the time from the start with people, on station, until completion of pumping and subsequent disconnection of the fill and drain lines. We weren't always trying to set new records though. The hardware and procedures had matured to that point, and there was always a great deal of luck involved in it, I thought. We started the launch countdown on September 9, 1966, but a pinhole leak in the Titan first-stage oxidizer tank caused us to scrub until after it was fixed the next day. It turned out that the Titan II weapon system had experienced these small leaks in their tanks after the missiles were deployed in their silos. The Martin Denver division had faced these problems before and solved them. But on Gemini, we had never encountered this problem. So if they had the problem on the earlier versions of the weapon system, there were probably guys at the Cape who had experienced it before. When we knew what the

---

[1]   Hacker, Barton C. and Grimwood, James M. *On the Shoulders of Titans*, NASA SP-4203, (US Government Printing Office, National Aeronautics and Space Administration, 1977), pp.354-357.

problem was, all the propulsion guys started saying, "We got to get Shakey Shelton back over here."

"Who is Shakey Shelton?" I asked. It turned out he was the one guy in the area who was a certified fixer of the leak. The leak was what we called a small fuzz leak at the dome weld on the first stage. Shakey was working at that time on Titan III, and he was made readily available with his water glass patch, which was the approved fix developed for the Titan II. He came over on the morning of September 9 and worked his magic. We had off-loaded the oxidizer before he got there. By the start of second shift, we were back in business for the next day's attempt.

Troubles with the target vehicle Atlas booster's autopilot on September 10 caused another scrub, and we didn't try to launch again until September 12. The countdowns for the two vehicles proceeded normally, and the launches were perfectly accomplished within the abnormally short launch window of just two seconds!

During the program, the Gemini spacecraft got gradually heavier, and the lifting capability of the Titan remained the same. With Martin under the terms of the incentive contract, the contract guys in Baltimore thought they should go on record with the concern about the weight increase. For each launch, a computer model that predicted the vehicle's performance was run, and it took into account variables such as predicted winds aloft, propellant temperatures, and loaded spacecraft weight. The computer gave the output in terms of payload weight apart from the ideal weight if all variables in the model were as expected. That is, if all variables were nominal, the payload margin would be neither negative nor positive. However, if the model indicated some variables nonnominal, then the payload margin would be negative. A negative payload margin of, say, 190 pounds meant that that many pounds would have to be taken off he vehicle to compensate for the nonnominal parameters in the model. Of course, there was nothing could be done at this point to correct the situation—it just meant that the mission would fly with a little less-than-ideal probability of making it. When the test conductor would poll the team members for a flight "GO", the incentive monitor would reply the Titan was "GO" with a negative payload weight of 190 pounds indicating some less-than-desirable amount of margin! He did this on the voice network that went everywhere! This shook me up when I first heard it, but I had heard it before this flight; and when he did it on my launch, I knew what to expect on that day. We never scrubbed a launch based on these measurements.

When the crew of Gemini XI reached orbit, they went through a series of orbital adjustments using their OAMS. Then, based on calculations they had performed, they burned the thrusters to rendezvous with the Agena. When they turned on their rendezvous radar, they were within fifty nautical miles of it. Eighty-five minutes after launch, on the first revolution (M=1), the Gemini met up with the Agena. A major program objective had just been achieved barely an hour after launch! After docking with the target, they tried several dock-undock procedures—all successfully.

The next day, the astronaut crew began preparation to attach the one-hundred-foot tether between the spacecraft and the Agena. Gordon had to exit the spacecraft in EVA gear and proceed to the docking adapter section on the Agena. In space, in zero gravity, and in a pressure suit, movement was very hard to control and very fatiguing for him. This had been the complaint of Gene Cernan on Gemini IX and of Collins on Gemini X—that everything took much longer and was more tiresome than anticipated, and that insufficient hand—and footholds had been provided. Zero gravity was hard to simulate on earth. Simulations on earth of the task that Gordon was now attempting were limited to a few seconds of zero-gravity environment created by the sudden diving of an airplane. The job of attaching the tether required a stance developed in training, with feet wedged between the docking adapter on the Agena and the recovery section on the spacecraft, astraddle the spacecraft (similar to one's position while riding a horse). In space, this technique wasn't working out as he had practiced, so on his second try he reached the target and had to hold on with one hand and use the other to attach the rope. With his legs about the spacecraft nose, holding on with one hand, fighting his space suit's tendency to propel him into space, he was able to secure the tether's clamp to the Agena after about six minutes of trying. "Ride 'em, cowboy!" shouted Conrad into his mike as Gordon was undergoing all these machinations. This task made him so tired he came back inside the cabin to rest before continuing the EVA. Aside from stowing things in the cabin and operating some experiments, the crew was done in for that day.

On the following day, the Agena was fired. This added approximately 297 meters per second (664 miles per hour) to their velocity. The spacecraft wound up 1,372 kilometers (852 miles) at its highest point above earth. From this point Conrad was able to discern the roundness of the earth, and he could see an arc of about 150 degrees of the earth's horizon.

Returning to operating altitude, Gordon did a stand-up EVA—one in which the outside hatch is open and the astronaut is standing in his seat, with his upper body outside in space while his lower body is being constrained inside the spacecraft. While in this position, they accomplished several hours of photography. This included photos of star fields as one of the experiments. The stand-up lasted until they crossed over Florida when Gordon announced that both he and Conrad had just taken a catnap during the stand-up EVA! Gordon was hanging on his tether while Conrad was inside.

After a rest period, the crew was ready to try the gravity experiments. They first attempted the gravity gradient experiment. The Agena at the end of the tether was unstable; and after waiting a few minutes for it to stabilize, the Houston flight controllers thought something must be wrong with the Agena's stabilization system, so they canceled this part of the experiment. Then the next part was implemented where the spacecraft was connected to the target via the tether and then spun up. The crew found that while trying to get the tether taut between the vehicles, the tether dynamics took a while to get damped out. Finally, Conrad beeped the thrusters to start the rotation, and the two vehicles at either end of the tether began to rotate around a point in the center in the tether at a rate of twenty-eight degrees per minute. Conrad then added more energy, and the rotation speed increased to fifty-five degrees per minute. The centrifugal force of this rotation created artificial gravity. Conrad and Gordon demonstrated it within the cockpit by allowing a camera to free float all the way from the instrument panel at the front of the spacecraft toward the rear of the spacecraft, on its own, under the influence of artificial gravity.

After demonstrating this artificial gravity experiment, the spacecraft separated from the target vehicle. Mission objectives then called for a rerendezvous with the target and demonstration of long-distance station keeping.[2]

Splashdown of the spacecraft occurred on September 15, 1966, after seventy-one hours in orbit. While getting ready for splashdown, the crew allowed the spacecraft to control the reentry automatically. Reentry had been a manual procedure, carried out by the pilot on previous missions. This was the first time the crew let the computer do it. The task included corrections to the flight path angle, adjustments for orbital altitude, cross-range corrections if required, and firing the retro-rockets. The Gemini computer handled the actions and events automatically and accurately. The position at splashdown was 4.6 kilometers (2.8 miles) from the recovery ship, USS *Guam*.

---

[2]   *Ibid.* pp.358-370.

Robert L. Adcock

Chapter 13

The Wrap-Up

The mystery of EVA was foremost in the minds of Gemini mission planners when the last launch was in the works. Although Gemini IV had been an unqualified success, EVA problems on Gemini IX, X, and XI were sufficiently complex that NASA rejected taking the USAF-developed AMU on Gemini XII, the last flight. In a letter from NASA to the US Air Force, Dr. Mueller explained that the last flight must be devoted to EVA fundamentals, and therefore the AMU would not be taken.

James Lovell and Buzz Aldrin were selected to fly Gemini XII, and Gene Cernan and Gordon Cooper served as alternates. Aldrin took on the challenge of understanding the fundamentals of EVA on this flight. Zero-gravity simulations on earth previously were practiced in an airplane. Each time the airplane dived down and pulled out of the dive at a steep angle, occupants inside the plane had a few seconds of weightless conditions. An entire day of flying the airplane could only provide a few minutes total of weightless flight. In 1966, weightless simulation in a pool of water was developed as a way of testing weightlessness. This overcame the difficulties and limitations of testing in the airplane. This helped Aldrin in his training for Gemini XII.[1]

---

[1]   Hacker, Barton C. and Grimwood, James M. *On the Shoulders of Titans*, NASA SP-4203, (US Government Printing Office, National Aeronautics and Space Administration, 1977), p.370-374.

As far as the launch vehicle was concerned, this was now getting to be old hat. Technical problems were few, the checkouts were going great, and we had plenty of time in the schedule, the spacecraft being the long pull (the one which took the most schedule time for preflight preparations). Markus Goodkind was assigned for this last launch, and I worked second shift some of the preflight.

Late in this last launch vehicle preflight preparation, we ran a Combined System Test. During the plus count, the time that simulated the flight, we got a spurious guidance signal that told the second-stage engine to shut down. This signal was termed sustainer engine cutoff (SECO). That was an error, but we didn't think it would be too hard to isolate the problem, so we relegated the fix to the guidance subsystem level guys. General Electric and the Martin guidance people worked several days without success, trying to replicate the spurious signal. NASA was justifiably concerned, and somehow the problem festered up to the point that NASA management got involved. I recall attending a meeting at KSC one afternoon, joined by all our electrical and interested spacecraft people, as well as NASA. I think Walt Kapyran chaired the meeting, and we were told to find the problem or else! We didn't want to go back into combined system test configuration because it would have invalidated some later operations we had already performed. We agreed to rerun the part of the test where the spurious signal came through. People began to be frustrated by the amount of time necessary to find this problem. Somebody dubbed the thing as, "Whatever Happened to Baby SECO?" It caught on, and everyone was talking about Baby SECO. The 1962 movie (*Whatever Happened to Baby Jane*) had just been released. Finally, one afternoon we were able to get another spurious SECO, and we isolated it without a doubt to the guidance package.

People who had elected to stay on with the program were naturally getting nervous about where they would be working after the last flight. Martin had indicated all Gemini labor would be surplus after the last flight. They would be paid their bonus money and then cut loose. Across the river at KSC, NASA contractors on the Apollo program were manning up, and some people went over and interviewed and got agreements for the contractors to wait until after our last launch

Robert L. Adcock

and then hire them. I know that Grumman hired a large number of our people for the Apollo lunar lander checkout and launch. North American Aviation also hired a number of our guys. They had the Apollo Command/Service Module, as well as the Saturn launch vehicle second stage that they produced.

I was naturally getting antsy about the employment situation myself and managed to get an interview with one of the North American managers. I told him on the phone that he would have to wait until our second shift started that day because I had to get it going before I felt safe to leave the launch complex. He graciously accepted this restraint, and I went over to see him in the NASA Launch Control Center at 6:00 PM. I had prepared a resume, and when I opened my briefcase to get it, out tumbled a novel I had been reading on the job! He correctly guessed that things were slow around Complex 19, and I had brought a book along just to help pass the time. Nevertheless, I was offered a job doing the same thing as on Gemini with little raise in pay. I rejected the job offer because I remember thinking I would lose my identity over there. I wouldn't be reporting to him, but one of his test conductors, who in turn reported to a test manager, who then reported to somebody, and so forth—you get the picture.

Sometime later, Joe Verlander, Martin's Gemini director, called me and said that John Coryell, Martin's Titan III director, had a job for me on Titan III; and he would call. Well, I waited around for a couple of days for his call and nothing. Then one day I said to myself that John was probably pretty busy, and I decided to call him. He said, "Sure, come on over here when you're done, and we'll put you to work." Happy news—I didn't have to change companies or anything else, just turn left and go to the ITL facility instead of right to Complex 19 when coming in to work and report to Roy Hunter. I already knew Roy and respected him. It would be all right.

On November 11, 1966, the Gemini XII countdown started. The astronauts came to pad 19 to enter the spacecraft with signs on their backs, one reading The and the other reading End. At 3:46 PM they launched into orbit. Their target vehicle preceded them into orbit, leaving the ground at 2:08 PM.

Rendezvous was first on the list to accomplish; and shortly, after little over an hour in orbit, the rendezvous radar locked on target. Shortly afterward, the data from the radar became intermittent to the point that the computer could not use it. The astronauts had to rely on manual observations and manual calculations of range-closure rates and distances from the target, together with charts on rendezvous, developed by Aldrin as part of his dissertation. Buzz Aldrin had become known as Mr. Rendezvous before the flight. After docking, a planned burn to achieve a high orbit was scrubbed when during its flight to orbit, the Agena engine had an unexpected 6 percent drop in thrust chamber pressure and turbine speed. Mission controllers did not trust the engine for a second firing. Since this part of the mission was dropped, time became available to photograph an eclipse of the sun that occurred during what would have been the Agena burn.

Next Buzz began the EVA maneuvers, commencing with opening the spacecraft hatch and standing up in his seat. He was very deliberate in all his tasks, studying each one for comparison with events later on. He limited the first phase to minor tasks involving replenishing camera film, installing a movie camera, and photographing star fields. This phase took two hours and twenty minutes. After a rest period, he moved outside the spacecraft and proceeded hand over hand toward the Agena, where he hooked up the tether to fasten the two vehicles together. Then he moved to the rear of the spacecraft area where he successfully tried out a new foot restraint called golden slippers. He then accomplished some experimental tasks, including torquing of bolts and cutting metal while in the rear of the spacecraft. Then he returned to the cabin and rested inside it and later exited the spacecraft and maneuvered toward the Agena. He opened an experimental box containing small kits for task accomplishment during the EVA. These included mating and demating of electrical connectors and using an Apollo torque wrench. During this episode, he was attached to the spacecraft with waist tethers that allowed him to use both hands in the tasks he undertook. He also tried out these same tasks while tied with one tether then while not tethered at all. On the way back to the cabin, Aldrin stopped and wiped the command pilot's spacecraft windows. Commander Lovell asked him to "check the oil." In a later

Robert L. Adcock

hatch opening, more EVA pictures were taken even though the picture-taking mission objectives had been completed.

Between these two EVA excursions, the crew set up to demonstrate the gravity gradient experiment. This meant separating from the Agena and getting in the vertical position, with the target vehicle on the bottom, vertical to the earth's surface, and the spacecraft on top. The one hundred-foot tether separated the two from each other. They had deployed the tether uneventfully, but attempts by the command pilot to draw the tether taut were unsuccessful although the establishment of the gradient was successful. In this configuration, the vehicles gyrated from time to time, it being reported that the spacecraft wigwagged over three hundred degrees at one time.[2]

Experiments took up the rest of the mission time line, and during its fifty-ninth revolution, the automatic reentry preparations were begun. Splashdown occurred only about three miles from the carrier *Wasp*. With splashdown, Gemini mission XII and the Gemini program ended.

---

[2]   *Ibid. pp. 374-381.*

Chapter 14

Everybody's Gone

What a letdown! Suddenly everybody was vanishing from the complex, and our old work procedures that we had so carefully developed became dead paper. People were scurrying to write final reports and get to their new jobs. Launch complex propellant loading systems—those having to do with propellant loading and pressurization—had to be made safe. We didn't get the postlaunch contractor back (the contractor who helped clean up the launch damage each time) after the last launch. No need for it! As I remember, two men, Jerry Walden and Vern Derby, stayed on the complex to tidy up things. I don't know how long they stayed. The cat Gemini who dwelled on the launchpad under the ramp, the pet of the propulsion and mechanical techs, had to find a new home. She had borne forty-nine kittens since she had been there.

When I think about the program, I can't help but reflect on the goals that were accomplished, not only for the space race, but also for the participants. Nobody left without a job, I'm sure, albeit with some other organization. Several left with a sizeable check in their pocket from the bonus the company had provided. Just about everybody got a more responsible job than the one they previously had. They got to cash in on their experience gained from Gemini. That was one thing I never thought of: all these people with manned-flight experience were going over to Kennedy Space Center to work on the most complicated space flight ever devised by man. Going to the moon has been rated as one of the humankind's greatest accomplishments. This infusion of manpower into organizations that put men on the moon must have aided the Apollo program immensely, although it's hard to measure. I know this was true for the NASA and civilian contractors, but at a recent Gemini fortieth reunion, it became obvious that some of the Air Force people had not been able to parlay their experiences into better jobs. The Vietnam War was going at that time, and some were reassigned to participate in that conflict.

The Gemini program developed techniques essential to going to the moon. The moon trip required rendezvous and docking as well as EVA. Astronauts who flew Gemini and gained so much flight experience became central to the astronaut corps for the Apollo program.

Everyone remembers the program vividly and fondly because it was a program where everyone gave their best. Charles W. Matthews, NASA's Gemini Program director said, "The thing that stands out in my own mind is the way in which the effort and dedication of many individuals and groups coalesced into an extremely effective team."[1]

We all loved and respected one another at the operations level. As I recall, no person got laid off unless he wanted it for some reason. We were so well educated in the art of manned space flight that jobs were readily available someplace on the Apollo program. But Apollo was so much larger than Gemini and under so much schedule pressure to land a man on the moon before the end of the decade that it took almost twenty times the number of employees in Florida that were on Gemini.

The mutual admiration that we all had for one another, I believe, is one factor that has helped the US Space Walk of Fame Foundation in Titusville, Florida, to germinate and grow. The organization was conceptualized in 1988 and sprang from an idea that Titusville medical doctor Chastain had when he saw the Hollywood Walk of Fame in Los Angeles. Nicknamed SWOF for short, it is dedicated to identifying and

---

[1] Hacker, Barton C. and Grimwood, James M. *On the Shoulders of Titans*, NASA SP-4203, (US Government Printing Office, National Aeronautics and Space Administration, 1977), p.vii.

honoring everyone who worked on the space programs, whether they were contractors, military or government employees. SWOF has erected three major monuments to date on the Indian River in Titusville, within sight of the Kennedy Space Center—one to the Mercury program and one to the Gemini and to Apollo. On each of these, the handprints of the astronauts who flew those missions are on the monuments in bronze. But the names of the not-so-well-known people who made the missions fly and got the crew safely back home are engraved for posterity in granite on the monuments.

In 1996, thirty years after the end of the Gemini program, I got a notice that the SWOF would be dedicating the Gemini monument in November 1997 and that astronaut John Young would break a bottle of champagne on the monument and make a speech. I wanted to see what these guys were doing, and I also wanted to protect my Gemini program memories because I felt I had some ownership in the program, so I attended. It is a magnificent monument and well up to the standards that I had in mind for such a program! I knew Cal Fowler, the president at that time, and he needed a SWOF treasurer. I volunteered, and it has provided me with so much pleasure for some of my retirement years. One factor is that SWOF maintains a name file database of everyone that is identified on the monuments. The SWOF organization fills the role for me, I believe, that the old country store used to provide where the men of the community, wanting to keep up with the local news, would gather at the store around a potbellied stove, play checkers, and swap gossip. Information on old friends and compatriots comes into SWOF's database, and it is a way of knowing where everyone is and what they may be doing. Visitors to the museum and gift shop also allow one to brag or tell it like it used to be, which is something that seems to be inbred in man.

We seem to have come full circle. Cal Fowler was a test conductor during planning for the launch of the Atlas Agena target vehicle from pad 14. Charlie Mars, who took over the USSWOFF presidency after Cal left, was a spacecraft engineer for KSC during the Gemini launches. I met both these guys back then, and now they are back in my life through association with the Space Walk of Fame.

Robert L. Adcock

In 2006, SWOF hosted a reunion for the Gemini program on the fortieth anniversary of the end of the program. Many of the original participants returned to tell war stories and visit old friends. Some had their personal memories (oral histories) recorded during the reunion days. Something about the Gemini program causes them to come back to review old times.

We are left with our memories of the Gemini Program. It was sandwiched in between Mercury and Apollo. Looking back, it is almost faded to black because our minds tend to discount long-ago events over those that stand out more clearly in the recent past. It is one purpose of this book, to not lose track of such an elegant program that was a necessary stepping-stone to our accomplishments in space.

Personally, my program experience had helped me with my career in aerospace. I went on to launch Titan III and the Vikings to Mars back in the 1970s. I concluded my tenure with aerospace in 1992. No doubt this was the most exciting program I was ever on! In the end, the Martin earned and was awarded a 100 percent award fee for their efforts!

Finally, we are reminded of the stalwart giants who, no matter whom they worked for, served the program in outstanding ways. Many have already been mentioned in this book elsewhere. Others include Air Force Chief Warrant Officer Bart Barton, the mayor of pad 19; H. H. Lutjen of McDonnell; Walt Williams of NASA; and General Richard C. Dineen, USAF. And who could forget the indelible impact that NASA-KSC's George Page had on the people and the program?

GLOSSARY

THE AEROSPACE CORPORATION
A private organization that gives the US Air Force technical advice and opinion.

AGENA
Agena was a rocket stage that was equipped to propel and stabilize itself in orbit, having been propelled there by an Atlas launch vehicle. In space, the Agena, outfitted with docking gear, could be used as the target vehicle so that the Gemini spacecraft could dock and fly with it. Agena propulsion was used to position the combination space vehicle in various orbits.

AGE
Aerospace ground equipment, used to check out and launch the Titan launch vehicle.

ASAP
As soon as possible

BOOSTER
The rocket used to get the payload to orbital velocity or to its destination.

BUILT-IN HOLD
A preplanned point in the countdown where a hold will occur—usually used to catch up with the work or to adjust the liftoff time to an exact moment.

CAPE CANAVERAL AIR FORCE STATION (CCAFS)
The land mass east of the Banana River called Cape Canaveral. This is differentiated from Kennedy Space Center lying directly west of CCAFS on Merritt Island.

CAPE
A shorthand reference to CCAFS.

CSAT
Combined System Acceptance Test, used to demo correct launch vehicle systems operation during vehicle sell-off.

EVA
Extravehicular activity, covers the activity of the astronaut in space outside his spacecraft.

FRR
Flight Readiness Review held by NASA a few days before entering into the terminal phases of launch preparations in which all progress to date is reviewed with upper management and the prognostication for the future is discussed.

GLV
Gemini launch vehicle, a Titan II derivative.

GYROS
Short term for gyroscopes used in the guidance systems for vehicle stability and direction.

GT-X
Gemini Titan, the space vehicle consisting of the Gemini spacecraft and the Titan launch vehicle, with the mission number following; thus GT-7 was Gemini Titan, mission 7.

HIGH RANGER
A bucket-type truck like those used by utility lineman except when it was extended, it was high enough to reach the spacecraft from the pad surface level.

HYPERGOLIC
The Titan propellants were hypergolic in that they would ignite when mixed together.

LAUNCH WINDOW
The span of time when the vehicle can be launched and still meet its flight objectives.

INCENTIVE FEE
The amount of profit a contractor is allowed to earn on a contract if he makes all his profit incentives. In the case of Martin, at minimum, they would receive costs incurred (no profit); but if they did well on schedule and performance, they would collect profits in excess of those generally awarded in a minimum-risk-type contract.

LIFTOFF
The moment the vehicle becomes airborne after ignition.

MCC
Mission Control Center, the flight controllers' home. First, it was located at Cape Canaveral and later at Houston, Texas.

MERCURY
America's first manned space vehicle carrying only one astronaut per flight.

NASA
The National Aeronautics and Space Administration.

OAMS
The Orbit Attitude and Maneuvering System on the Gemini spacecraft. Used to change orbital planes, attitude and generally maneuver the spacecraft in orbit.

PAN AM
One of the range contractors, Pan American World Airways.

RCA
The other range contractor, Radio Corporation of America.

READY ROOM
A building located outside the fenced Launch Complex 19 which served as offices and conference center.

SPACECRAFT
The Gemini module that went into orbit and in which the flight crew lived, survived, conducted experiments, and which delivered them to their destinations in space or on earth.

SPLASHDOWN
The time when the returning flight hit the water in the recovery area.

TEST PROCEDURES
Written procedures used for testing or operating either airborne or ground systems or both in the preflight preparation. For the launch vehicle there were on the order of one hundred procedures performed before and during each launch cycle.

THROW WEIGHT
The amount of payload weight a booster can lift to orbit.

USAF
United States Air Force. There were two organizations: the US Systems Command, headquartered in Los Angeles who contracted for the hardware and bought the contractor services, and the 6555th ATW,

located at Cape Canaveral, which was the operational arm of the systems command.

USSR
Union of Soviet Socialist Republics.

VTF
Vertical Test Fixture at Martin's Baltimore plant.

Vignettes

How I Was Hired at Glen L. Martin

Right out of college, I had taken a job at ARO Inc. who ran the US Air Force Arnold Engineering Development Center at Tullahoma, Tennessee. The AEDC was an installation of wind tunnels that ranged from subsonic to hypersonic, giving the Air Force the capability to test airplane and rocket models over a wide range of speeds. It was all brand new when I went there with my electrical engineering degree straight out of the University of Tennessee in 1952. ARO, the installation operator, gave us a week's orientation before I was assigned to the power control division. The work consisted of things I had never heard about, like checking the insulations condition on all large insulators and internal insulation of large electric motors. The only thing palatable about that job was that the Air Force was accepting the property from the US Army Corps of Engineers, and they asked ARO to help test the installations. This allowed us to come up with some unique tests to accept the installations, which was kind of interesting. But as far as our general role at that installation, I thought it was boring, kind of like watching paint dry, going from one machine to another, repeating the same insulation tests run the year before, and recording the results.

ARO was a new company and had their share of problems, like a consistent wage scale across the board that took into account experience and education. Frequently the technicians made more than the engineers they were working with. Pretty soon everybody in power control knew what everybody else was earning and when their next raise was due! It didn't make for a happy family. Still, I lasted three years, hoping they would wake up to the great employee they had and make things right (according to my definition of right)!

One day in 1955, a friend of mine came over and showed me an ad in the *Nashville Tennessean* that asked for engineers to interview the next day for possible jobs in Baltimore at the Glen L. Martin Company. It sounded good to me; and the next day, I got up and got ready for work, but instead I drove to Nashville and went to the hotel where the job interview was being held. I got there early and checked in with the Martin guy from the desk, and he told me to come on up.

So we met and exchanged pleasantries, and I thought, *Boy, this guy is really hung over.* His eyes were red, and he had that sleepy look. He asked me to tell him what I had been doing, and as I began to drone on about my boring job, I looked up and he was asleep! I stopped talking at that point, and he immediately awoke from his nap and offered me an all-expense-paid round-trip to Baltimore for other interviews. I often wonder what would have happened had he not slid into that snooze that day!

I went to Baltimore the next week and met my future boss for an interview. He liked me, and I signed on with Martin and commenced on May 31, 1955. I lived there until 1956 when I transferred to Florida on Project Vanguard, the Navy's program for orbiting an International Geophysical Year satellite. The IGY was designated as an eighteen-month-long period beginning in July 1957 and ending December 31, 1958. When we tried to orbit the first American satellite in January 1958 the first American satellite, we struck out when the Vanguard crashed and burned on the launch pad. Later, Vanguard met and exceeded its contractual commitment by putting three satellites in orbit.

In October 1957, the Russians orbited Sputnik. We on Vanguard were ticked off! We could hardly believe they had accomplished what we were assigned on Vanguard to do!

What was the purpose of the Russian thing? we wanted to know. Something that flew overhead and beeped! My roommate at the time was a young Martin engineer who had come to Florida from Baltimore to work on Vanguard propulsion. His name was Robert Webster Stone, and we knew a young Navy guy who invited us to hangar S on Saturday night following the Sputnik launch to hear the Sputnik traverse over the Cape. Bob and I got dressed and went back to *work* in hangar S, in time to hear its beep, beeping on its way overhead that night around 10:00 PM. This was the start of the space race.

The Americans wanted to know how the Russians could beat us putting up a satellite and one that was so much heavier than anything we had planned for in the near future. The heavyweight (184 pounds) attested to the larger size of the Russian booster rocket that was needed to orbit Sputnik over existing American-built rockets. The longer it stewed at the national level, the more finger-pointing until at last the race for space began, in earnest. We had to catch up, and this set the stage for going to the moon on Apollo.

My Soldering Skills

On the Vanguard program, our techs and engineers had wired and plumbed the entire launch complex when we first came on to Launch Complex 18A. In late 1956, that was my job—to get the electrical control wires installed and working. I spent a lot of time working with the techs and learning techniques they used. We got into the countdown on the first missile from that pad. It was my first countdown. We were going for a static firing—one in which the first-stage engine would be turned on but the missile itself would be held down. In this way, all the ground equipment would be checked against the airborne missile, and any problems would be manifested.

Well, sure enough, we made it to T-5 minutes. At that point, the vehicle was put on internal power, and the engine igniter had a continuity check with the ground power source. I was not on console; I just heard our boss say we had a problem that needed to be checked out before we went further. I finally got someone to explain what had happened, and the two of us went to the pad. The ordnance guy disconnected the igniter while I, guessing at the problem, got some tools and a meter. We looked at the three-wire cable that went to the igniter cable, measured it, and found it had been cross wired!

On that day in October, the Florida weather had suddenly gone from a balmy 85 degrees to the mid-60s; and when we went to the pad, I had no windbreaker, just short sleeves. Up on that test stand the wind was blowing 20 mph, and overhead, the LOX tank was spewing and venting (they often make pops and cracking sounds when full of the 275 degree liquid); I was cold and nervous. Why I didn't get one of my techs to come help, I'll never know. I took a Weller soldering gun and touched it to the cable pins, and I was shaking so bad, I bent the tip 90 degrees. But it kept working, and we finished resoldering the cable under the worst of conditions, plugged it back in, and resumed the count. My fist experience underfire I guess!

Working at Martin

Martin paid their employees well, judging from my own satisfaction with my compensation. When I first joined Martin in 1955, I got an enormous pay raise over what I was making at ARO Inc., the operator of the Arnold Engineering Development Center at Tullahoma, Tennessee. Martin paid me a whopping $600 per month to move to Baltimore and live. Most college graduates were drawing close to $4,000 a year in those days!

When I first arrived at Baltimore, I was assigned to the B-57 Air Force bomber program. They assigned me to a drawing board, and I began struggling with the Martin drawing system, trying to make sense out of two hundred designers in one room, turning out blueprints for an airplane that was already flying! Huh? It turned out the Air Force had many versions of the B-57, and all these guys were working on an advanced version, the B-57 E. This was a tow-target airplane—one that could tow a gun target for shooting practice and also serve as the bomber it was designed to be. The tow-target version had not flown yet.

So pretty soon I began to talk to the guys sitting on my left and my right, both of whom could outdraw me. Their finished drawings were so nice and clean, and mine were smudged with eraser marks! I knew nothing about airplanes! I didn't know a fuselage from an empennage! I also found out what they were making moneywise, and I was shocked. Their salaries seemed like a pittance compared to mine. I hated being on the drafting board, and I wasn't good at it. I considered quitting, but I figured if Martin could pay me what they were paying and yet hire

Robert L. Adcock

drafting talent like these compatriots of mine sitting at my elbows for so much less money, then my assignment would change soon!

Next I was assigned to a proposal effort in a different part of the plant. It was a satellite program we proposed to build. At least I was off the drafting board and had a desk. One day I noticed that other people sitting on the proposal, working at their desks, would open the top left-hand drawer, bend over, and then close the drawer again. I looked more closely and found out they all had a coffee cup in the drawer. In Martin, you could not drink coffee at your desk or anyplace else except the cafeteria. It seems that during World War II, the employees abused the coffee break, turning it from a ten-minute break to over a half hour by the time they got back to their machines. Never mind, we had nothing to do with the shop anyway! I never knew I liked coffee until I found out it was not allowed.

One day out of the blue, as I sat there trying to come up with something unique to put in that proposal, a guy came up to me and said, "You got a raise!" Huh? I asked somebody else whether that guy was for real or not. I didn't recall ever having seen him before. Welcome to the world of matrix organizations! It turned out his name was Harry Creamer, and he was a real sponsor of mine, but at that time I was too dumb to know you needed a sponsor! Although not on the organization charts, nevertheless just about everybody agreed that having a sponsor (someone in an influential position in the ranks who liked you) was a good thing and was really required for your own good.

Early Cape Days

When I first came down to the Cape, I began to hear over the Capewide public address system the superintendent of Range Operations running other tests on other missile programs, and I fell in love with the authority and the obvious responsibility he exercised. In those days the SROs would announce the status of the test over the PA system so that those with a responsibility on that test could know the status without having to check personally with the SRO. The organizational man on our program who had that kind of authority and responsibility was the test conductor. Test conductors, Capewide, had a kind of mythical quality about them that most admired. They were in the thick of things where decisions were made and where information flowed freely upon which to base these decisions. With that kind of reverence and awe associated with the job, I knew I wanted to be a test conductor someday. One day on the Vanguard program, while I was still the electrical lead, Stan Welch and I walked the rocket down from top to bottom. As we were passing about the middle level, there were some propulsion guys with a water hose, washing down a fluid spill, I guess. All I could see was how wet the inside of the rocket was getting, which meant my electrical wires were also getting drenched and therefore degraded. I couldn't help myself—I cussed and told them in no uncertain terms to knock it off! Welch liked my reaction so much he told me he would like me to be in operations as a test conductor.

I got my chance to launch one on the last Vanguard shot. It was a night launch, and at about T-30 minutes, the range searchlight operators would turn searchlights on the vehicle and light up the entire test stand area where the vehicle was setting. This was primarily for camera-tracking purposes, yet there may have been a safety consideration also. In any case, it was bright as day out there on the rocket. Nothing was automatic in a Vanguard launch. Consoles were lined up according to the functions they provided during the launch operation. The test conductor stood behind the console operators and looked over their shoulders. The most important of these consoles was for the electrical control and firing. There was also a propellant loading and pressurization console at which the operator, late in the count, was trying to keep the liquid oxygen (LOX) tank full because LOX boils at minus 273 degrees. In a rocket tank, it must be resupplied at all times, to the last minute. So as test conductor, I was overlooking the guys operating the consoles when I looked at the ground pad in the searchlights and something was spilling from the LOX tank vent all the way to the ground, like liquid! So I said to the operator, "What the—is that?" He replied he put too much LOX in and had run the tank over, and I was one scared man. I didn't know if that would hurt anything or not. He didn't seem upset. He said we should ignore it, and I told him not to do that again. By this time the count was getting low, and we closed the vent valve and pressurized the tank, which took care of the problem.

Robert L. Adcock

On Vanguard, when the countdown reached zero, if NASA yelled fire, then the firing console operator activated a switch that sent the signal, starting the engine, to the stage 1 engine igniters. That night Vanguard put up a thirty-some-odd-pound satellite—the heaviest of the program to date.

Friday Nights all over America

I make no bones about it—Friday nights were sacred to me. From the get-go I had a friend, Dave Mackey, who felt much the same way. In the rocket business, it was customary to stop by your favorite watering hole on the way home from work and talk over the week with guys who worked on different programs. You might even have two or three drinks! Dave would say, "It's Friday night all over America!"

Everybody at the Cape had work problems. Some were common to everyone, like the early phone system. You couldn't call long distance (if you did, they couldn't hear you, or you them). Traffic to and from the Cape was slow, and the food there was less than tasty in the Pan Am cafeteria. Contractor and government employees used to get together for a drink and to talk about the unusual events in their lives that week. We were beginning to find out several of our problems were similar and may have had common solutions. Anyway, the Surf and Ramon's in Cocoa Beach were wise to us, and they had "two for ones" on Fridays, and maybe you would stay for dinner too! The Surf would give away a free drink to anyone in the bar when a missile was launched at the Cape. Chances were that missile had traveled along the unpaved street A1A from Patrick AFB to the Cape before launch because early-launch facilities at the Cape were slow in coming.

Communications from the Cape to anywhere else was a problem. Telephones didn't work—besides that, they figured you didn't need to talk to your wife anyway because you might let out a secret launch date or the time in the countdown that was going on for that particular day. All launch times were secret in the early days, and many a launch *suddenly* sprang to life from the Cape, surprising a number of veteran watchers living down in the beach towns. You could tell the veteran watchers from the novices—they did not mistake the lighthouse for the missile that was counting down for a night launch. Sometimes the novices were left standing on Cocoa Beach while the veteran watchers headed back to their homes when the missile scrubbed. Veterans began to figure subtle ways of communicating the countdown time to their families so they could watch from the beach with minimum exposure to the mosquitoes!

Range operations had good communications to all the blockhouses, downrange stations, and intrarange sites. Not only did they have the MOPS (Missile Operations Phone System) system, which afforded the user with voice-to-voice communications over any one of dozens of channels, but also, in the case of downrange stations, they had ship-to-shore radio, as well as all kinds of radar. The Range broadcast, over a Capewide public address system installed in all the hangars and pads, the countdown status of any major test involving Range support. The superintendent of Range Operations (SRO) would announce something like, "Test 2532, T minus thirty-five minutes and counting." All you had to know is who test 2532 was, and of course, if it was your launch test, you would know.

One unique set of phones the SRO had was called green phones because they were painted green, and they went from point to point. For example, the SRO had a green phone from his desk to the pad safety officer console in the blockhouse. When he needed to talk to Pad Safety at that particular pad, he simply picked up the green phone and it buzzed on the pad safety desk in the blockhouse. Sometimes green-phone users would forget to hang up after a conversation, and this disabled the green phone, so sometimes you would hear over the PA system words like, "Pad Safety, blockhouse 18, hang up your green phone."

When the veteran missile watchers became more proficient at finding out about the launch times and gathering at their favorite watching spots along the beaches, it sometimes seemed that there was direct communications between the bars in Cocoa Beach and Range Operations. As a gag, I heard more than once an announcement over the Capewide PA system, "Bernard's Surf, hang up your green phone!" Of course, there was no green phone in Bernard's Surf, the Cocoa Beach restaurant—where someone seemed to be hanging out all the

Robert L. Adcock

time. People just figured out how to communicate secret information without getting caught. Range and contractor management later on became very stern about anyone revealing any a priori launch information, although adding NASA programs to the Cape launch manifests caused the launch information to be more available. The NASA programs' launch information usually was open and published, which was antithetical to the Department of Defense philosophy of secrecy and stealth.

Later on, when Gemini came along, other bars were in on the offerings. Some were elegant, like Ramon's, atop the glass bank building in Cocoa Beach from which the whole downtown area could be viewed. Some were bizarre, like the Vanguard Lounge at the end of Highway 520 at the ocean edge in Cocoa Beach where, from the bar and unbeknownst to the swimmers, a side view of the pool from underwater could be seen by those drinking at the bar. This had the flavor of peeping through a knothole into the girls' shower room when you were in high school.

At the Cape, for food, Pan Am offered the roach coach, the only source of food delivered to the blockhouse while you were in countdown. The roach coach started out early in the morning and drove from launchpad to launchpad or wherever working people were. It had two people on it, and on this particular one, a young black woman, Marilyn, drove the van; and a motherly-looking gray-haired elderly lady, Hortense, fried hamburgers as they went down the road. They would pull up to the blockhouse and feed everyone. People used to announce, "The roach coach is on the approach." When in countdown, there were always a lot of guys from out of town—visiting, helping with the test—who came up to the wagon to get food. One day, some guys murmured that they hoped the food quality exceeded the looks of the servers to which Hortense quipped, "We're here to feed you, not to look good!"

Postlaunch parties were the thing even if the launch failed, but especially if it made it and especially if it was a new kind of missile that hadn't flown before. In 1958 the first US Air Force Thor intermediate-range missile flew from Cape Canaveral pad 17. It flew on Friday, and the postlaunch party started in Cocoa Beach thereafter. I passed through Cocoa Beach about noon the next day; and at the intersection of Atlantic Avenue (A1A) and the Minutemen Causeway, just as I rode along, people began spilling out of the bar across from the Surf and began hula hooping in the middle of the unpaved intersection! Evidently, the party still had legs.

Baltimore Colts

In the fall of 1966, Martin's Management Club, which met every month for dinner and fellowship at the Patrick Air Force Base Officers' Club, had sought out and acquired for one particular meeting Don Shula and his Baltimore Colts football team. They were on a trip to Miami; they stopped for a tour of the Gemini Launch Complex 19. I got to meet Don Shula. What a wonderful, warm, pleasant personality he has, I thought. There were in all about ten people together, and I don't remember the players; they were superstars, I am sure. They took a tour of the launch vehicle spacecraft and blockhouse, and they chatted with us as they were leaving. They seemed to have little understanding of what they saw, and I am sure whoever got them there appealed to them on the basis that the Gemini launch vehicle was made in the same town where they lived and worked.

Lyndon B. Johnson

One day we were notified that the President of the United States would tour the Gemini launch complex. His visit was off-limits to Martin or other contractors but of course vested in NASA and, because of their arrangements with DOD, to some degree with the Air Force. We cleaned up the place and got it shining, ready to receive the President right after lunch. At the same time we were cleaning the hardware, the FBI and others were clearing the place from the personal safety point of view. This resulted in us having to evacuate some of our blockhouse desk areas and stand by maybe one hundred feet away from the blockhouse entrance. Air Force One, the President's airplane, was supposed to land at the skid strip and discharge the party and proceed directly to Complex 19. Well, we continued to wait outside the blockhouse, trading rumor after rumor that Air Force One had been spotted while time dragged on. Suddenly, around 3:00 PM, there was a flurry of activity outside the main gate leading up to the blockhouse as a mighty sedan roared through the gate entrance. Out jumped Lyndon Johnson surrounded by young guys in black suits who led the way. My thoughts at the time were that this man was large; he had on a black suit in this torrid weather. He obviously had just freshened himself up before getting off the plane; he had just shaved and powdered up. His heavy beard showed through the fresh white powder on his face. He was talking to someone, which caused him to bend low because of the relative stature of the two. I wondered whether this man even knew where he was or what was being done there; he was so busy whispering in that man's ear! Suddenly they were out of the blockhouse back into the auto and out the gate. I don't think he even went to the launchpad!

Index

D

Denver, Colorado, 23, 26, 52
DynaSoar, 23, 24, 25, 26

E

engines, 15, 17, 18, 25, 26, 36, 40, 46,
    57, 85, 86, 87, 101
erector, 15, 16, 19, 42, 43, 46, 49, 50,
    55, 56, 85
EVA. *See* extravehicular activities
extravehicular activities, 9, 11, 20, 21,
    91, 96, 108, 109, 111, 114

F

Feagan, Wally, 16, 19, 43
Flight Readiness Review, 51, 52
fourteen-day mission, 86, 88

G

Gannett News, 13
Gemini
    launch vehicle, 13, 26, 27, 80, 84
    launch vehicles, 48
    objectives, 9, 11, 79, 88, 89, 90, 92, 104
    Program, 118, 119
    program, 9, 11, 27, 33, 50
    space vehicle, 11
    spacecraft, 14, 15, 79, 80, 82, 93, 94,
        99, 101, 104, 107
Gemini 2, 56
Gemini I, 62
Gemini II, 25
Gemini III, 21, 61, 62, 100
Gemini IV, 9, 11, 13, 19, 20, 21, 91, 111
Gemini IX, 108, 111
Gemini V, 64, 68
Gemini VI, 80, 81, 82, 83, 84, 86, 88,
    90, 92
Gemini VII, 80, 81, 82, 83, 84, 86,
    88, 95
Gemini VIII, 90, 91, 96, 99

Gemini X, 99, 101, 102, 108
Gemini XI, 104, 105, 108
Gemini XII, 111, 113
General Electric, 10, 26, 39, 112
GLV, 29, 30, 31, 33, 34, 54
Gordon, Richard, Jr., 92, 104, 108, 109
gravity gradient experiment, 105,
    109, 115
Grissom, Virgil I., 11, 61
Guadalcanal, 102
Guam, 110

H

hard start, 59, 90
heat shield, 59, 63
Hello, Bastian C., 27, 28, 30
Honeywell Inc., 26
Houghton, Jim, 15, 16, 46
hurricane, 55, 56, 68
hypergolic fuels, 17

I

inertial guidance system, 26
intercontinental ballistic missile
    (ICBM). *See* Titan II

K

Kapyran, Walt, 39, 112
Kennedy Space Center, 32, 67, 85, 90,
    100, 112, 116, 118
Kennedy, John F., 40
Kraft, Christopher, 16, 64, 65, 66, 67
KSC. *See* Kennedy Space Center

L

LaFrance, Jerry, 27
Launch Complex 16, 100
Launch Complex 19, 11, 13, 33, 39,
    54, 55, 61, 100, 113
launch operations, 33, 45, 51, 87
Launch Operations, 19, 36, 47, 81, 88,
    92, 132

Robert L. Adcock

launch window, 106, 107
launchpad, 11, 15, 18, 27, 33, 34, 37,
    39, 40, 41, 42, 47, 48, 52, 54, 56, 79,
    80, 81, 82, 84, 85, 86, 87, 92, 100,
    116, 128, 135, 137
launchpad 14, 79, 91, 92, 118
Leonard Mason, 94
Leonov, Aleksey A., 20
lightning strike, 54
Lockheed Missiles and Space
    Company, 10
Lovell, James, Jr., 20, 83, 84, 86, 96, 99,
    111, 114

M

malfunction detection system, 26, 59
Manned Space Center, 64, 65
Martin Company, 13, 19, 24, 25, 33,
    34, 41, 44, 48, 56, 60, 61, 64, 92,
    100, 106, 119, 127, 130
Martin, Glenn L., 44
Matthews, Charles, 88, 117
McDivitt, Jim, 11, 13, 15, 20
McDonnell, 14, 15, 25, 44, 60, 92, 95
McMechen, Ed, 52
Mercury
    program, 11
Mission Control Center, 21
Mueller, George, 88, 111

N

NASA, 14, 21, 37, 39, 48, 50, 57, 62,
    68, 82, 84, 85, 87, 88, 94, 96, 97,
    104, 111, 117, 133, 135
    Flight Control, 21
    Launch Control Center, 113
    Launch Operations Center, 32
    Mission Control, 49, 79
    Operations, 65
    National Aeronautics and Space
        Administration. *See* NASA
North American Aviation, 113

O

OAMS, 20, 62, 93, 94, 99, 108
Orbital Altitude Maneuvering
    System. *See* OAMS
oxidizer, 86, 90, 106, 107
oxygen, 50, 66, 91, 132

P

PAA. *See* Pan American
    World Airways
Page, George, 57
Pan American World Airways, 35, 36
Patrick Air Force Base, 133, 136
pilot safety, 10
pogo, 40
pogo modification, 40

Q

quality assurance, 14, 19, 52

R

Radio Corporation of America, 35, 36
Radio Frequency Systems, 46, 50
Range contractors, 37, 87
Range safety officer, 36
RCA. *See* Radio Corporation of
    America
rendezvous, 9, 11, 20, 21, 50, 62, 67,
    79, 80, 81, 83, 84, 88, 90, 91, 92, 93,
    96, 97, 99, 100, 101, 102, 105, 106,
    108, 110, 114, 117

S

Schirra, Walter M., Jr., 83, 85, 86, 88
Schlechter, Bob, 23, 24, 27, 29, 47
Scott, David, 92
Sequence Compatibility Firing, 40
Space System Division, 33
Sputnik, 128
SRO, 36, 37, 38, 131, 134

21558070R00076

Made in the USA
Lexington, KY
18 March 2013